The Universe

Paul Murdin

The Universe
A Biography

First published in the United Kingdom in 2022 by
Thames & Hudson Ltd, 181A High Holborn, London WC1V 7QX

First published in the United States of America in 2022 by
Thames & Hudson Inc., 500 Fifth Avenue, New York, New York 10110

The Universe: A Biography © 2022 Thames & Hudson Ltd, London
Text © 2022 Paul Murdin

Typeset by Mark Bracey

British Library Cataloguing-in-Publication Data
A catalogue record for this book is available from the British Library

Library of Congress Control Number 2021943328

ISBN 978-0-500-02464-5

Printed in China by Shenzhen Reliance Printing Co. Ltd

Be the first to know about our new releases,
exclusive content and author events by visiting
thamesandhudson.com
thamesandhudsonusa.com
thamesandhudson.com.au

Contents

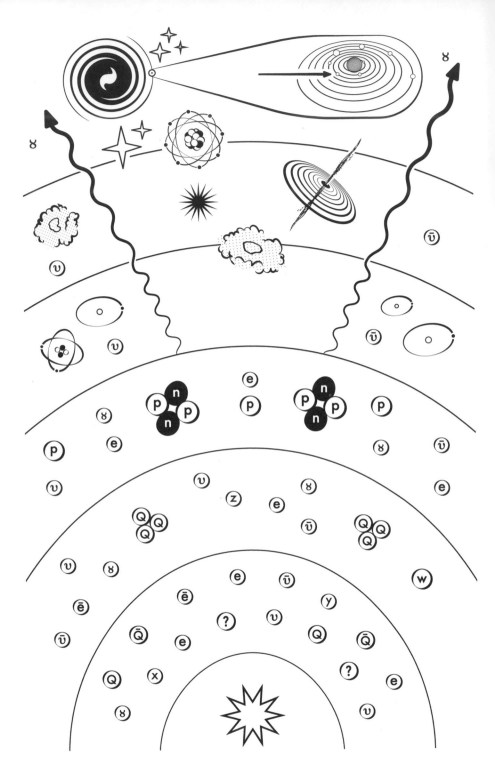

From the birth of the Universe as exotic fundamental particles, its biography continues as matter and dark matter in a dance of stars and galaxies, gathered into clusters.

Preface

The word 'astronomy' can mean both the activity through which astronomers discover what happens in the Universe and the science itself, in which the discoveries are described. A history of astronomy is usually a narrative about the first – the way that the science of astronomy developed throughout human history, from early times to the present. Although some of this history is recounted here, this book is primarily the story of the Universe – its birth and its growth – even if, as with the life of a historical figure, some chapters of its biography are obscure or missing.

The book paints a picture of what has happened in the Universe, starting from its beginning in the first milliseconds of the Big Bang expansion. In the Prequel (Chapter 13), I have also described some of the things that may have happened even before the Big Bang, and in a Sequel (Chapter 12) some of the events that are likely to happen in the future.

Like biographies in general, this one is broadly chronological, although sometimes it has made more sense to describe what happened in overlapping sequences – I have sought to write an understandable story rather than one that is strictly chronological. A time marker at the top of each right-hand page locates when the events described in the text of that sub-section took place.

Astronomy is, in its essence, the science of the very large, even though small interactions in its tiniest components – molecules, atoms and subatomic particles – cause its most dramatic events. As a result, despite it being the weakest of the forces, gravity plays more of a role in astronomy than in other sciences because it has great effects on large masses, like planets, stars and galaxies, even if they are separated by large distances. The Earth and the minor bodies of the solar system, such as meteorites, are the smallest astronomical bodies that I consider.

This is my selection of what I believe are the most important events, the greatest structures, the most powerful explosions, the largest ecosystems and the celestial bodies in the life of the Universe that are most relevant to our human story.

There are lots of billions in this book – billions of stars, billions of galaxies, billions of light years, and more. The word is formally not used in science, but in this book 1 billion is equivalent to 1,000 million.

1

The Questions That Revealed the Universe was Born

The Universe was born about 13.8 billion years ago. If it had not been born, and if, therefore, it has existed for ever, this book could not have been written as a biography. There would have been no progression or development of the Universe over time and everything would always remain the same. But, gazing into the far distances of space, astronomers can see changes laid out in a timeline and this book is an attempt to put what they see into words, like a biography.

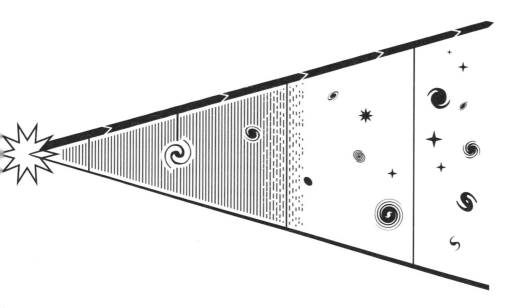

In the Big Bang, a soup of fundamental particles condensed into the matter that we see today, and dark matter that hides from us. After a period of darkness, stars and galaxies emerged into the Cosmic Dawn.

Why is the sky dark at night?

The dramatic birth of the Universe was a small, dense, hot explosion, the fireball of which is still visible as radiation everywhere. Galaxies condensed out of the outrushing material, a phenomenon that we see as the expansion of the Universe. Somewhere in that event and that material lie our own origins.

There is a simple fact that justifies the analogy between the history of the Universe and a human life, which ages after starting at birth: it is dark at night. In daytime, when we look up, our line of sight goes up to the sky. It may zoom straight out into space, but it may reach an air molecule and its direction may then be changed. The line of sight might end up on the Sun's surface – it forms a route via the sky that links our eye and the Sun. As a result, the sky is bright. At night, our line of sight may also zoom directly into space; again, it may reach air molecules and be diverted, but the molecules will not be illuminated by sunlight. The line of sight will not then turn towards the Sun, but will extend out into space to end up somewhere far away in the Universe. Sometimes it will end up on the surface of a galaxy or star, but mostly it heads into nothing and, as a result, the night sky is dark.

If the Universe was infinite in extent and fully populated with stars, the line of sight would always end up on a star. If you stand in a large forest, surrounded by trees, no matter in which direction you look, eventually your line of sight ends up on a tree trunk. Likewise, in an infinite Universe populated by stars, your sight-line would always end up on the surface of a star and the night sky would be as bright as the surface of the Sun. Manifestly this is not so.

This contradiction is known as Olbers' paradox, after the early nineteenth-century German astronomer Heinrich Wilhelm Olbers (1758–1840), who was not only a prominent doctor in Bremen, but also a keen amateur astronomer. As a student, he studied both medicine and mathematics, and it was said that he developed a new way to calculate the orbits of comets while sitting at the bedside of a sick patient. He installed a telescope in an upper room of his house, from which he observed comets. He survived on only four hours sleep at night and was thus able to pursue two careers: a busy professional life as a doctor and his passion as an astronomer. He has been called the greatest of amateur astronomers (and his work as a doctor seems to have been respected, too).

Bremen is a port in northwest Germany between Denmark and the Netherlands, near to the North Sea. Its climate is not ideal for astronomy, and it might have been while he waited for the clouds to clear that Olbers wrote an influential article in 1823, still important and the subject of much discussion in modern times, on the paradox of why the night sky is dark. His article brought attention to the paradox, although the question has a much older history, with a number of distinguished scientists having discussed it before him. Its importance as a help in appreciating the birth of the Universe was not rediscovered until 1960.

Olbers formulated the paradox as if the Universe was more or less uniformly populated by an infinite galaxy of stars, which was the common belief at that time. We now know that our Galaxy only extends to a distance of about 200,000 light years, but the Universe is more or less uniformly populated by galaxies out to a much further distance; the change from stars to galaxies does not change the essentials of Olbers' argument. The solution to the paradox is that there must be long gaps between galaxies, sight-lines that pick their way through the galaxies like corridors so that we can see through the gaps to the empty region that is beyond them all. This is as if the trees, in the analogy above, were actually grouped in a small wood and between some tree trunks we could look out into the open countryside beyond.

What Olbers' paradox implies is that at night, through the gaps that make sight-lines through the galaxies, we can see to the boundary of the Universe. There are no galaxies or stars beyond that boundary. So, the majority of sight-lines head towards nothing at all and that is why it is dark at night. This explanation is couched in terms that are easy to visualize, and the essence of it is true enough, but not the detail, because the Universe is not a collection of galaxies isolated in otherwise empty space. It is a curved and finite region that in its total has a large volume entirely filled with galaxies.

Olbers was limited by not knowing about the curvature of space because he lived a century before Albert Einstein, but he put together the important argument that the Universe is limited in its size. This has an even more significant consequence – the Universe must therefore also be limited in time because the limitation in size is set by its age. As our line of sight extends into the distance, it also penetrates back into the past because light travels at a certain speed. We see the image of the past carried to us by that light. Because the Universe was born, no light can reach us from a time before its birth – the age of the Universe sets a horizon in our view of space, beyond which we cannot see. The boundary of the Universe is at a distance corresponding to the distance that light has travelled since the birth of the Universe. The darkness of the night sky, as an observation interpreted with the speed of light in mind, compellingly puts forward the proposition that the Universe was born.

In general, the entire biography of the Universe is laid out backwards along a line of sight through space from here on Earth outwards – out to a certain, limited distance and back to a certain, limited time. Astronomers can witness the sequence of events over the time since the Universe was born by looking into the distance. In principle, if astronomers can view the entire Universe, they can view its entire lifetime. Of course, the earlier events are further away and less distinct than the nearby ones, so the more distant history is less clear than the nearer history. But that is true of all history. Moreover, the events witnessed out there in the past are not the *actual* predecessors of the events taking place nearby now. They are, however, events *like* the predecessors of events nearby.

The goal of having the lifetime of the Universe laid out like a timeline in large part explains why astronomers are obsessed with building ever larger telescopes. Everyone knows astronomers use telescopes to peer into distant space, but they also use them as machines to look back in time. The bigger the telescope, the further it can see, not only into space but also into time past (pl. II).

Why doesn't the Universe collapse?
The Universe is a collection of galaxies spread out over space. They all attract each other by the force of gravity. Finding the solution to the obvious question – why do the galaxies not pile up into a big heap in the middle? – eventually led to the discovery that the Universe is expanding and therefore must have a starting point.

This question of collapse is one that worried the English physicist Isaac Newton (1642–1727), who discovered the force of gravity and realized that it was a force by which everything attracted everything else, no matter the distance by which they were separated. The story is that in 1665–66 an outbreak of the plague began to spread from London to cities and villages elsewhere in Britain, including Cambridge, where the young Newton was studying at Trinity College. The university locked down to reduce the exposure of its students and teachers to the epidemic. Newton left his rooms in college and returned to his home to self-isolate in a bubble with his family in the country, on a farm in Woolsthorpe in Lincolnshire, not far geographically from Cambridge but far enough from the contagion in the city. There he had time to think. According to an account given about 1727–28 by John Conduitt, a colleague and relative by marriage, Newton described the event that, at the age of twenty-three, inspired his thoughts on gravity:

> In the year [1666] he retired again from Cambridge on account of the plague to his mother in Lincolnshire & whilst he was musing in a garden it came into his thought that the same power of gravity (which made an apple fall from the tree to the ground) was not limited to a certain distance from the earth but must extend much farther than was usually thought – Why not as high as the Moon said he to himself & if so that must influence her motion & perhaps retain her in her orbit, whereupon he fell a calculating what would be the effect of that supposition.

The apple tree, or more likely a descendant of the tree, is still there in Woolsthorpe, outside the farmhouse door.

Newton had realized that gravity might be a force that pervaded – indeed, dominated – the entire Universe. It kept the Moon in its orbit around the Earth, the planets in their orbits around the Sun – and presumably the stars in orbits around each other. It was a big thought to have been provoked by a falling apple, even if Conduitt's story is not entirely reliable as a historical account, perhaps much improved by being retold repeatedly by Newton himself as he grew older. The French journalist Voltaire popularized the story of the falling apple and the Moon, which he had learnt from Conduitt himself, as a successful everyday image that conveys Newton's ideas about the universal attraction of gravity.

In Newton's greatest work, a book first published in 1687 and known as *The Principia*, he identified how the force of gravity between two attracting objects depends not only on their masses (the more massive each one is, the greater the force) but also on the distance separating them (the gravitational force lessens as objects get farther apart, diminishing according to the square of the distance between them). This is called the inverse square law.

Newton soon realized that his concept of gravitation created a problem for the sustainability of the Universe. If the Universe was some sort of container filled with (as he thought) stars (we would say galaxies), it would be unstable. It would soon collapse, gathered together by the mutual attraction of everything for everything else. Newton exchanged correspondence on the subject with Richard Bentley, a theologian with strong scientific interests, a controversial and tyrannical Master of Trinity College, Cambridge, with whom Newton formed an alliance (it was Bentley who took *The Principia* through the Cambridge University press). Newton wrote in a letter to Bentley in 1692:

> As to your first Query, it seems to me that if the matter of our Sun & Planets & all the matter in the Universe was eavenly scattered throughout all the heavens, & every particle had

an innate gravity towards all the rest & the whole space throughout which this matter was scattered, was but finite: the matter on the outside of this space would by its gravity tend towards all the matter on the inside & by consequence fall down to the middle of the whole space & there compose one great spherical mass.

Newton went on to sketch out one of his most prescient speculations, which we will see later came to describe the origin of galaxies in the outrushing material of the Big Bang:

But if the matter was eavenly diffused through an infinite space, it would never convene into one mass but some of it convene into one mass & some into another so as to make an infinite number of great masses scattered at great distances from one another throughout all that infinite space.

Today we call this process 'gravitational collapse', and it is a concept that, as we will see, is one of the most important ways that cosmic history developed.

Newton's law of gravitation had other problems that made it difficult to understand – gravitational collapse was only one. Think for a moment about the inherent implausibility of the proposition that a force can be transmitted from one body to another through space, through nothing at all. Nevertheless, Newton's law was amazingly successful in describing the motion of the planets, so it seemed to be right, or, at least, to do the right things. Even now, Newton's discoveries are used to control the orbits of spacecraft through the solar system from one planet to another. Newton turned aside from the difficulties with his theory of gravity with the famous remark *Hypotheses non fingo* ('I make no hypotheses'), published in some reflections that he added in the second (1713) edition of the *Principia*. He took an empirical view: his theory worked, even if he did not understand everything about it.

The difficult issue of why the Universe did not collapse was marked as an open question for two hundred years. Then in 1916, German-born physicist Albert Einstein (1879–1955) published his general theory of relativity, which was essentially a new theory of gravity, a refinement of Newton's. However, when applied to the entire Universe, it produced the same result as Newton's theory: the Universe collapsed. Einstein was bolder than Newton: he did make a hypothesis. He suggested in 1917 that there might be a kind of negative gravity that propped the Universe up, so that it was stable and could last indefinitely. Like Newton, Einstein did not say exactly what the force was, but instead worked out some of what would result if that was so, introducing an entirely empirical factor into his equations symbolized by the Greek capital letter lambda (Λ), which he called the cosmological constant.

Einstein's solution to the problem of gravitational collapse was built on some incomplete mathematics, as was shown by the Russian mathematician Alexander Friedmann (see page 41). Clever though he was, Einstein had failed to map out all the possible outcomes of his own general theory. There was a way forward that avoided the outcome that the Universe collapsed and did not last forever without requiring the prop of the imagined cosmological constant Λ. Indeed, the new idea would have worked for Newton, too. The correction gave birth to the idea of an expanding universe. At first Einstein rejected the correction, but he came to accept it, and then reject it again. In reminiscing about these developments, he later came to regard his own idea about the prop as the 'biggest blunder he ever made in his life'. For about a century afterwards, cosmologists worked out their theories of cosmology using the cosmological constant Λ, and then set $\Lambda = 0$. It was not until the 1990s that cosmologists in general came to believe that Λ was not zero after all (see Chapter 2).

The primeval atom: Lemaître's model of an expanding Universe

It was Belgian priest and astronomer Georges Lemaître (1894–1966) who, in a series of papers written about 1930, brought the mathematical theory into a physical picture of the Big Bang. He studied engineering and fought heroically in the First World War as an artillery officer. After the war had ended (and perhaps in reaction to his experiences), he turned to more peaceful studies, both as a Jesuit priest and as a mathematician, pursuing an interest in astronomy and cosmology. In 1923, he studied at the University of Cambridge under the astronomer Sir Arthur Stanley Eddington (1882–1944). Eddington had been one of the first people to see the importance of Einstein's general theory of relativity and encouraged Lemaître to discover more of what general relativity had to say about the life of the Universe.

Lemaître pursued Eddington's suggestions when he returned to his own university in Leuven in Belgium, discovering how Einstein's theory allowed the possibility, overlooked by Einstein himself, that the Universe was expanding. This implied that in the future it would become more and more rarefied and, effectively, end.

Neither Einstein nor Eddington liked this idea, both heavily influenced by a religious feeling, that, once it had been created, the Universe would last forever. This was thought to be most consistent with the biblical account of the Creation as recounted in the book of Genesis, and in the teachings of the Jewish, Christian and Islamic religions that God is boundless. Eddington and Einstein were both religious people, one a Quaker and the other a free-thinking Jew, and each man found repugnant the idea that the Universe was not eternal.

One might have thought that Lemaître, a Roman Catholic abbé, would have had a similar strong opinion about this question. However, he was a follower of Saint Thomas Aquinas (see page 270), and said: 'It appeared to me that there were two paths to truth. I decided to follow both of them.' Lemaître let his reason direct him to cosmological truth, whatever it was, and then to reconcile that aspect of the truth with his theological beliefs. The Big Bang emerged as Lemaître's favourite explanation by science and reason of the origin of the Universe, giving a clear picture that the Universe started and will continue indefinitely. He was able to get to the same idea by following the path of faith as mapped out in the book of Genesis in the Old Testament, a canon of holy books accepted by all three of the Abrahamic religions.

Lemaître came to his concept of the expanding Universe through abstract mathematics but he provided a very specific interpretation in a form that was readily imaginable. If the Universe was now expanding, it evidently started off in a more concentrated form. He envisaged this starting point as a primeval atom that exploded. Lemaître thought of this atom as a dense assembly of all the atomic particles in the Universe, which, like a radioactive element, spontaneously disintegrated, setting off the life of the Universe. As the originator of the concept of an explosive start to the Universe and as a Belgian abbé, Lemaître doubly earned the honour of the title 'Monseigneur Big Bang'.

The idea of a primeval atom was couched in terms of the cutting-edge science of the first part of the twentieth century. From the time of the Greek philosophers, atoms had been regarded as the fundamental particles of the Universe – the very word 'atom' is derived from the Greek, meaning 'something that cannot be divided'. It is an immediate challenge for a scientist to hear someone say that they have got to the bottom of a problem, its ultimate cause or its fundamental origin; the scientist's instinct is to ask why this is the beginning of the story, why there is not something behind the 'ultimate' reason.

In the early years of the twentieth century, scientists probed successfully into atoms and found that they are in fact composite, made of electrons, protons and neutrons. Each atom has a nucleus of protons and neutrons surrounded by clouds of electrons, which can be envisaged, in some ways, as being in orbits that step outwards from the nucleus, like a small solar system of planets orbiting their sun. For years these were regarded as *the* fundamental particles, and this was the way that Lemaître, who knew about the structure of atoms, envisaged his primeval atom. His Big Bang was an explosion of an accumulation of all the electrons, protons and neutrons in the Universe.

In the latter half of the twentieth century, even protons and neutrons have been discovered to have structure, and the phrase 'fundamental particles' now encompasses over three dozen distinct elementary particles. I name them to give an impression of their variety and the complexity of the science of the Big Bang explosion at this early stage: there are six quarks (with the names up, down, strange, charm, bottom, top), six antiquarks, six leptons (electron, muon and the tau particle, and their respective neutrinos), six antileptons, thirteen gauge bosons (eight gluons, the photon, the W+, W-, Z particles and the graviton) and one Higgs boson (see Chapter 13 for explanations of some of these particles) – all derived from something that was initially thought to be indivisible! The structure of these so-called fundamental particles is even now under discussion with theories that use entities called strings to build them up.

Fundamental particles are studied in terrestrial laboratories by generating them in an atomic reactor or by colliding particles in high-energy accelerators like the Large Hadron Collider at CERN near Geneva. In some ways, the Big Bang was a similar reactor or high-energy accelerator, but it was not disciplined by laboratory control and generated the entire range of possible particles. The ambitious goal of cosmologists is to start at the most basic components of matter, whether thought to be a single primeval atom, a collection of atoms, a plasma of fundamental particles or a spectrum of strings, and to write the biography of the Universe from that moment.

Has the Universe expanded?
Lemaître's concept of the birth of the Universe in what came to be known as the Big Bang received immediate proof in a discovery in 1929 by the American astronomer Edwin Hubble (1889–1953). Hubble was a clever man, educated at the universities of Chicago and Oxford in mathematics, astronomy and jurisprudence, who thought he would take up a career in the practice of law. After the end of the First World War, however, he 'chucked the law for astronomy' (his own words) and began work at the Mount Wilson Observatory in California.

Hubble teamed up with astronomer Milton Humason (1891–1972), a man who started his association with the Mount Wilson Observatory by driving a mule train up the mountain with components for the 100-inch Hooker Telescope being built there. Having participated in the telescope's construction, Humason stayed working in the same place, at first as a janitor and then as a scientific assistant. In this role he grew to understand better than anyone else how to coax the best out of the temperamental telescope. With Humason's help with what was then the largest telescope in the world, Hubble gathered data about a number of galaxies and found a way to estimate their distance. He went on to look at the motions of the galaxies using measurements taken by astronomer Vesto Slipher (see page 258) at the Lowell Observatory in Arizona, and in 1929 discovered how they were moving. The galaxies were all receding from us with speeds that increased as their distance increased: the trend line is now called Hubble's Law.

The natural interpretation of Hubble's discovery is that the Universe of galaxies is expanding. If indeed this is so, and we live in a typical galaxy, we would see all the galaxies around us receding, with the more distant galaxies receding faster. Hubble had seen the evidence that the Universe had indeed exploded, as Lemaître had visualized.

Hubble's interpretation of the evidence was that it confirmed the Big Bang theory and this was validated by further investigations. If there had been an explosion, the galaxies must have been packed tighter together in the past. This question could be investigated through the phenomenon of 'look-back'. If we look at the most distant galaxies, we are seeing them as they were in the past and we can investigate whether they are packed closer together or not. Up to the 1950s, the distances to which optical telescopes could see were not so great that they could see the difference in density, but radio telescopes were discovering thousands of radio-emitting galaxies at large distances and the question could be addressed by radio astronomers.

The new technique of radio astronomy developed from the wartime technology of radar. When the scientists and engineers who worked on radar in the Second World War returned home to civilian life, several of them grouped into universities and research institutes to carry out such studies in Britain, America and elsewhere. In Britain, the organizations now known as the Nuffield Radio Astronomy Observatory of the University of Manchester at Jodrell Bank near Chester, and the Mullard Radio Astronomy Observatory of the University of Cambridge at Lord's Bridge near that city, had their origins in this time.

It was appropriate that a new technique applied by newly created scientific groups would address the dramatic question of whether the Universe expanded or not. It was not an easy one to answer. Because the technology of radio telescopes was new, it was not properly understood and the results from different groups were inconsistent. Radio astronomers took up two sides over the issue. One camp was led by Martin Ryle (1918–1984) at the University of Cambridge; the other camp was a loose alliance of research groups in Australia. The cosmological question boiled down to a matter of counting the radio-emitting galaxies discovered by each group, but to count them you have to discover them. Ryle invented a new method to do this called aperture synthesis; in the countryside close to Cambridge, near a disused railway station, he used the straight and level railway line as a foundation to build a radio telescope to survey the sky. After the first experimental surveys, in 1955 Ryle published a catalogue of nearly two thousand radio sources called 2C (the second Cambridge catalogue of radio galaxies). It clearly showed a large overabundance of faint radio sources. Broadly speaking, 'faint' implies 'far' and 'far' implies 'old', so the conclusion was that the Universe was denser in the past. This favoured the Big Bang theory.

Although the Cambridge conclusion was correct, it was based on false evidence. Australian surveys showed a slight excess of faint sources but it was nothing like the excess found by Ryle. The Australians suggested that most of the faint sources in the 2C catalogue were spurious instrumental effects, not, in fact, real. In the *Australian Journal of Physics* in 1957, they did not shirk from a clear conclusion: 'there is a striking disagreement between the two catalogues...discrepancies, in the main, reflect errors in the Cambridge catalogue and accordingly deductions of cosmological interest derived from its analysis are without foundation'. They were right. Ryle redoubled his efforts to find the real faint sources.

Eventually, the cosmological arguments were resolved through the new work, which produced reliable catalogues called 3C (see page 72) and, later, 4C. There was indeed an excess of faint radio sources that were at vast distances and look-back times. Galaxies were, in general, packed closer together in the past and radio astronomers had confirmed that the Universe has indeed expanded. Ryle was awarded the Nobel Prize in Physics in 1974 for 'pioneering research in radio astrophysics...for his observations and inventions'. It is interesting that the prize was given for inventing the techniques, not for the cosmological discovery itself.

Since this work, there has been general consensus that the life of the Universe started with the Big Bang. Powerful modern optical telescopes like NASA's Hubble Space Telescope have looked back and pictured some of the innumerable galaxies 90 per cent of the way back to the Big Bang or more and seen how they crowd together.

From the Big Bang into the future
Lemaître had an interesting idea of how the life of the Universe would unfold. It came to produce you and me, he thought, in a way that might not have been determinate. He based his idea on the then newly developing theory of quantum mechanics. Like atoms today, the primeval atom would have existed in some atomic configuration. It exploded from that state, the change governed by quantum mechanics. Quantum mechanics includes the uncertainty principle, according to which we do not know precisely how things will turn out until they do. The Austrian physicist Erwin Schrödinger (1887–1961) explained this principle by posing the famous question of the cat in a box (known as Schrödinger's cat): when a cat is in

the box with the lid closed, is it alive or dead? When we open the box, we know it is one or the other. But before that, inside the box, the cat is both alive *and* dead, indeterminate.

In an article in *Nature* in 1931, Lemaître expressed his thoughts on the way the explosion of the primeval atom led to ourselves in terms of the obsolescent technology of a phonograph (gramophone) and the vinyl discs that it plays: 'The whole story of the world need not have been written down in the first quantum like a song on the disc of a phonograph. The whole mass of the world must have been present at the beginning, but the story it has to tell may be written step by step.'

We are a part of the history of the Universe and our origins lie in its birth. The history was indeterminate and might or might not have led to where we are today – we might have been alive or dead. We are in fact alive and we can trace the general steps by which the Big Bang created the environment in which we live, but there is no predestined path from the Big Bang to us.

Although Lemaître's primeval atom is now regarded as a scientific metaphor, not a literal truth, the basic idea that the Universe was born from the explosion of an energetic, dense accumulation of subatomic particles has survived to the present day. The next page of the life of the Universe sets off from this thought towards what was, at the beginning, our uncertain future.

2

The Big Bang: The Birth of It All

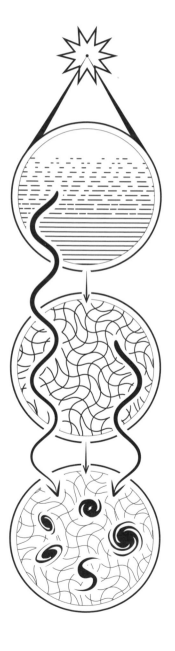

The Universe started to explode as hot, dense material made up of fundamental particles. Our comprehension of just how those particles interacted has developed with advances in scientific understanding of atomic, nuclear and particle physics since the last century. The first recognizably modern notions were inspired by what happened in much smaller explosions – those of atomic bombs.

At birth, the Universe rapidly inflated and exploded, creating gravitational radiation. Later, the explosion cooled and released the Cosmic Microwave Background radiation, showing the embryonic patterns in which galaxies formed.

Ylem: the formation of matter

George Gamow (1904–1968) was a Russian-American theoretical physicist, educated in Russia in the 1920s at the time that the theories of relativity and quantum mechanics were being developed. As a university student he became interested in nuclear physics and tried more than once to leave the poverty-stricken and oppressive life in the Soviet Union. On one occasion he set out with his wife to paddle across the Black Sea from the Crimea to Turkey in a kayak made of rubber stretched on a frame of sticks. They carried food for five days, including strawberries and two bottles of brandy, indicating Gamow's joyful attitude to life, his daring confidence and his profound, and in this case unfulfilled, optimism. On the second day, a storm arose and they were blown back onto shore, requiring them to concoct a story to cover up their escape attempt.

In 1933, as a respected Russian intellectual, Gamow was ordered by the Soviet authorities to attend a scientific conference in Brussels and he managed to arrange for his wife to accompany him. They both defected and moved through a succession of temporary academic jobs in France and England, before ending up in the USA. Gamow applied his knowledge of nuclear physics to astrophysics and then, during the Second World War, to nuclear explosions. Like Georges Lemaître, he turned away from his employment in warlike activities towards cosmology and brought together these various kinds of physics into the study of the explosion of Lemaître's primeval atom in the Big Bang (see Chapter 1). He worked with a number of distinguished scientists, becoming influential both in the work that he did himself and in the way that he affected the wider development of cosmology.

Gamow's most famous work on the nature of the Big Bang was carried out through his association with his American student Ralph Alpher (1921–2007), with the participation of American scientist Robert Herman (1914–1997): in the 1940s they all worked in various capacities at what was then a US government defence laboratory, the Applied Physics Laboratory in Maryland, a suburb of Washington DC. Alpher and Herman were employed to work on proximity fuses for torpedoes and antiaircraft gunfire; Gamow was a consultant on explosions. By its association with defence, their work on the Big Bang took place behind incongruously guarded gates.

The three of them started out by imagining that for the first milliseconds of its existence, the Universe was a dense maelstrom of incredibly hot fundamental particles bathed in hot, energetic radiation. It expanded and grew. The particles moved quickly and collided fiercely, interacting, changing dynamically from one to another. The characteristics of the mixture of particles altered progressively as the explosion expanded, cooled and became less dense.

Alpher called this material *ylem*, a term derived from a Latin word used by medieval theologians for the primordial substance from which everything was made. In modern terms, ylem was a hot plasma of fundamental particles: quarks, gluons, electrons and neutrinos (see Chapter 13).

Dark matter

There was another constituent of the material of the Universe that was made in the Big Bang; we know far less about it although it was crucial to the way the Universe developed immediately afterwards, as well as in the way the Universe behaves to the present day. That component is known as dark matter.

Dark matter is similar to ordinary matter in the respect that it generates and responds to the force of gravity in the same way. In interactions between particles of matter, energy is often released in the form of radiation or light, so if dark matter does not interact, it therefore generates no light. This is why it is called 'dark': it betrays little sign of its existence apart from its effects through the force of gravity. There is more dark matter than ordinary matter, so it is very important in ruling the Universe through gravity, but, except by gravity, it interacts very little indeed with ordinary matter.

The gravitational effect of dark matter on ordinary matter – stars and galaxies – is the way that it was discovered. The first indications came in 1933 through the work of Swiss-American astronomer Fritz Zwicky (1898–1974) at the California Institute of Technology (Caltech) in Pasadena. Zwicky was a man who was known for both his originality of thought and his irascible manner. He studied the largest agglomerations of matter in the Universe, namely clusters of galaxies.

Galaxies agglomerate into groups ranging in numbers from a few to many thousands. A group of a few galaxies can best be described as galaxies in orbit through and around each other. The larger clusters of galaxies number thousands and are abuzz like a swarm of bees. Zwicky examined such a cluster of galaxies in the constellation of Coma. He looked at the light from the stars of the galaxies and estimated the amount of mass that he could see. He also determined how fast the galaxies were moving and estimated the mass whose gravitational attraction would make them move that quickly. He found that the motions of the galaxies implied that their mass was four hundred times greater than expected from their luminosity, meaning that most of the matter in the cluster must be dark.

At first, the dark matter was thought to be dark stars, like black holes or neutron stars, but attempts to identify the precise type of star were negative. Most astronomers formed the easier conclusion that there was something wrong with Zwicky's work, whose conclusions were controversial. Zwicky did not readily accept criticism, however, and rebuffed his critics.

As it turned out, Zwicky was right about dark matter in principle, although the numbers that he derived in 1933 have had to be updated. His estimate that there is four hundred times more dark matter than luminous matter turned out to be a considerable over-estimate, even allowing for what is now known as the intra-cluster medium (hot gas revealed by its emission of X-rays, which was invisible to Zwicky at the time, there being no X-ray telescopes). Nevertheless, eventually, his general conclusion that there is much more dark matter than luminous matter in the Universe became accepted. Some of the key confirmatory work was carried out by Gamow's student Vera Rubin, working at the Carnegie Institute in Washington DC.

American astronomer Vera Rubin (1928–2016) also had a career that included difficult relationships with some co-workers, although her problems lay in a completely different area from Zwicky's. Naturally curious and observant as a child, she was educated at Vassar College in Poughkeepsie, New York State, then a college for women only where the concept of women astronomers was unremarkable, but after completing her education she found it hard to carry out research on equal terms with men. She struggled to break into the top level of astronomy, dominated by men as it was in the USA in the 1960s. She chose a specialism that seemed to be relatively uncontroversial, where she could get on with her work with less gender politics to fight. She investigated galaxies and their motions by use of spectroscopy and homed in on the problem of how galaxies rotate. Little did she realize that this would propel her into a deep controversy of the sort that she had wanted to avoid.

Working out how galaxies rotate had become a problem with the development of radio astronomy in the 1950s. When radio telescopes recorded the hydrogen gas moving in galaxies, radio astronomers found that generally the gas was moving too quickly compared with what would be expected from the mass that the galaxies contained. However, the interpretation of the radio observations was confused by limitations in the technology of radio telescopes, which were able to look at the entire galaxy all at once but were not able to compare the motions of individual parts of the galaxy (taking part in the rotation of a galaxy, the stars on one side of a galaxy move towards the Earth and stars on the opposite side move away and thus can reveal the rotational speed and period of the galaxy). Optical observations do not have as much of a limitation as radio telescopes and Rubin decided to investigate the problem by using a spectrograph (a spectroscope that records a spectrum) on large optical telescopes – a suitable tool to use because it can measure the motion of stars in different parts of a galaxy.

Rubin's investigation took on new capability when she teamed up with a talented maker of spectrographs, Kent Ford. His spectrograph was many times more sensitive than previously attained and could investigate the motions of stars in the very faint outskirts of spiral galaxies, the parts furthest from the centre of rotation. Rubin overcame the problems she encountered in gaining access, as a woman, to the observatories to which she could take the spectrograph and make the investigations she thought necessary. She recounted how she found that one observatory had no female toilets, so she commandeered a male toilet by cutting a skirt from a piece of black paper to stick onto the trousered male silhouette on the door to change its designation.

To her surprise she found that the stars in a typical spiral galaxy (see page 77) moved much faster than she would have anticipated based on her estimates of the mass of the galaxy. This was particularly true of the stars in the most distant reaches, where it would have been thought that, the gravity of the galaxy having been diminished by the inverse square law (see page 14), the stars would be moving slowest. In fact, they were typically moving at the same speed as, or even faster than, the stars further in. The stars in general, particularly the outermost ones, were moving so fast that they should have been flying off into space rather than circling their galaxy. The mass of the galaxy had to be much greater than she had estimated in order to keep the stars bound in their orbits, so there seemed to be something wrong with her calculation.

Rubin had estimated the galaxy's mass by measuring the light that came from its stars, then calculating back to the mass of the stars that could produce that light. If there was more mass in the galaxy than that, it was not stars – it was dark.

Since Rubin's work, the techniques of radio astronomy have improved and radio telescopes have been able to study the rotation of hydrogen gas clouds within spiral galaxies in their outermost regions where there are no stars at all, just tenuous gas moving with the stars in orbit around the galaxies. This confirmed and extended Rubin's conclusion.

There was more of a contradiction than Rubin thought: light from the stars diminished with distance from the centre of the galaxy, but the rotation of many galaxies did not reduce at all. There must be proportionately more dark matter in the outermost parts of a galaxy, compared to the mass in the form of stars: a typical galaxy of stars is embedded in an outer halo of dark matter.

There is typically at least five times more mass in dark matter in the Universe than in stars and gas. This is not so dramatic a conclusion as Zwicky's estimate of four hundred times, but a factor of five is still a large measure of our ignorance about the constituents of the Universe. The best modern guess about the composition of dark matter is that it comprises individual particles known as WIMPs (weakly interacting massive particles) or axions. It is thought that the particles were made in the Big Bang and, like the hydrogen made at the same time, persist even now. For dark matter to have the effects on the motion of galaxies that are calculated, there must be enormous numbers of these particles; perhaps 100 million of them pass through each of us every second. We do not notice this, so they truly must be 'weakly interacting'.

WIMPs and axions were not conjured up by cosmologists out of the blue: they had been conjectured to solve some other problems in particle physics. They have never been successfully made or detected in particle physics machines, even the most powerful ones like the Large Hadron Collider at CERN in Switzerland, so their properties are almost entirely enigmatic, but, although their very existence is not proven, most physicists and astronomers want to believe in them, or something like them, to solve otherwise intractable problems in astronomy and cosmology.

Neutrinos in the cosmic fireball

Quarks came together to make the ordinary material of the Universe so that in the first second of its existence it consisted of neutrons and atoms of hydrogen in a bath of photons and neutrinos. The hydrogen atoms were in fact split apart into their constituents, their orbital electrons freed from their nuclei. Each hydrogen atom was split into a proton and an electron, floating freely. The interactions of the neutrons, protons and electrons contributed to the vast numbers of neutrinos in the first second of the cosmic fireball.

Neutrinos are tiny, electrically neutral particles that feature in nuclear reactions of various kinds; radioactive decay is the kind in which neutrinos were first found. When the nucleus of a radioactive element like radium decays, it emits an electron, which carries off energy in a process called beta decay. It was natural for physicists studying beta decay in 1911 to try to measure the energetics of this process as a clue to how it happens. This revealed that energy was also being carried off by another particle, of a type not identified until then.

The two who made the fundamental discovery that led to the recognition of neutrinos were German chemist Otto Hahn (1879–1968) and Austrian-born physicist Lise Meitner (1878–1968), working in Berlin in 1911. They were both members of German physicist Max Planck's research group. Hahn had a straightforwardly brilliant and influential scientific career, including the award of a Nobel Prize. Meitner's career was more challenging. She joined the group from Vienna where she had grown up as a member of a large Jewish family; she was educated privately because, as a woman, she was not allowed to attend a university. She went on to become a senior, although under-supported, scientist in the Berlin group. After her work on beta decay, she and Hahn turned to work on nuclear fission. It was for this work that Hahn was later awarded the Nobel Prize in Chemistry in 1944, but not Meitner.

Although Meitner had converted from Judaism to Christianity and was an Austrian citizen, her position in Nazi Germany became progressively untenable and in 1938 she fled to Sweden. Looking back on her life in Berlin, she was highly critical of those German scientists, her former colleagues including Hahn, who, effectively, collaborated with the Nazi regime if only by remaining silent; she did not exempt herself from the criticism. By contrast, Hahn was praised by Albert Einstein as 'one of the very few who stood upright and did the best he could in these years of evil'. Having discovered and revealed the science of nuclear fission, Hahn was certainly tortured by the destruction and loss of life caused by the use of atomic weapons in Japan in 1945 and assumed the blame on his own shoulders. Looking back at this period of history, I am happy not to have been tested by being stretched on these ethical racks.

Meitner's life was one of achievement, in spite of everything that conspired to make her life very difficult. In 1911, she and Otto Hahn found that the energy of electrons emitted in beta decay was typically less than the energy that the radioactive nucleus had lost. It seemed that the reaction violated the principle of the conservation of energy – energy seemed to have disappeared. Where had the lost energy gone?

The conundrum was solved in 1930 by Austrian physicist Wolfgang Pauli (1900–1958), then working in Switzerland. In a daring hypothesis, Pauli suggested that radioactive decay must emit not only an electron, but also an undetected second particle, which carried off the missing energy. Looking at all the other constraints, Pauli concluded that the particle had to be electrically neutral and (almost) massless. Pauli was aware of how outrageous this claim was, saying: 'I have done a terrible thing, I have postulated a particle that cannot be detected.' Nevertheless, he was correct. Even if he had not made other even more important discoveries in quantum mechanics, for which he was given the Nobel Prize in Physics in 1945, it seems likely that it would have been given for this work.

The newly identified particle was electrically neutral like a neutron. Discussing the new particle with fellow Italian physicist Enrico Fermi, physicist Edoardo Amaldi coined the name *neutrino* (Italian for 'little neutron'); Fermi gave the word currency. The neutrino was eventually detected in 1953 by American physicists Clyde Cowan and Frederick Reines by studying the particles emitted from the Savannah River nuclear reactor in South Carolina. Reines was belatedly awarded the Nobel Prize in Physics for this discovery in 1995, but so long had passed before the importance of the discovery was recognized that Cowan had died before the award was made, so he missed out (under the rules, a Nobel Prize cannot be awarded posthumously).

It is now known that there are three kinds (or 'flavours') of neutrino. They transmute spontaneously from one flavour to another in what are known as neutrino oscillations. Neutrinos scarcely interact with matter at all; one estimate for the neutrinos made in nuclear reactions in the Sun is that they would on average travel through light years of solid lead before being absorbed by it, so travelling through billions of light years of empty space is not a problem for them. Even in the incredibly dense Big Bang material, neutrinos were liberated about 1 second after the Big Bang. Once made and liberated, these neutrinos have journeyed through the Universe without being absorbed and survive even up to the present. There are calculated to be about 300 Big Bang neutrinos in every cubic centimetre (0.06 cubic inch) of space in our part of the Universe at this time.

It would be important to detect some of these ancient neutrinos because they are some of the oldest messengers about the Big Bang, carrying information from when it was about 1 second old. An experiment at Princeton University in the USA called PTOLEMY is being developed to try to do this. The acronym has been contrived for the clumsy name 'Princeton Tritium Observatory for Light, Early-Universe, Massive-Neutrino Yield' and refers to the Greek astronomer of the second century CE who lived in Alexandria in Egypt and promulgated what was – until Copernicus proposed the heliocentric theory in 1543 (see page 37) – the most widely accepted geocentric theory of the solar system, or the Universe as it was then understood. The acronym effectively lays the claim that the neutrinos that it hopes to detect will be similarly foundational for modern cosmology.

The first chemical elements
In his doctoral thesis, Alpher studied the properties of what he and Gamow called ylem. In 1946, he discovered that the elements could be made by nuclear fusion, the same process that occurs in a hydrogen bomb and in stars like the Sun. In that process, Alpher envisaged that neutrons are successively bonded together to build heavier and heavier elements. This could happen for as long as the Universe was both hot enough and dense enough – it had to be hot

because heat made the neutrons move quickly and collide violently, and it had to be dense so that collisions took place frequently.

Gamow and Alpher wrote a paper describing this. Gamow was so struck by the sequencing of the process as neutrons built up elements one by one that he invited another physicist, Hans Bethe, to join the authorship of the paper, so that the authorship would be cited as Alpher, Bethe and Gamow (alpha, beta and gamma are α, β and γ, the first three letters in sequence of the Greek alphabet). Gamow added a playful footnote to the paper saying that Bethe was an author *in absentia*, giving the game away that Bethe was not an author at all. However, the footnote was omitted in the publication process. Bethe later regretted being added to a scientific paper on which he had done no work and that he afterwards had reservations about.

Alpher and Gamow thought that the fusion process would continue right down the list of chemical elements, building up quantities of all of them. This, they believed, was the origin of all the elements – all of chemistry would be traceable to the first few minutes of the Universe, in the effects of the Big Bang on its constituents. However, they were working with incomplete knowledge of nuclear physics and the characteristics of the Big Bang material. As a result of more detailed analysis, scientists concluded that, after less than a quarter of an hour, the temperature and density of the Universe had fallen to the point where nuclear processes stopped. The Universe would have subsequently expanded an enormous amount and cooled the radiation bathing the matter of the Universe to a low temperature. Gamow and Herman calculated that the temperature at the present time should be 5 degrees Kelvin above absolute zero. It was a good estimate: the radiation was detected in modern times and its temperature measured to be 2.725 degrees Kelvin.

The building of the elements in the Big Bang did not get much past lithium (element 3), with hydrogen by far the most abundant (element 1; 96 per cent) and helium in second place (element 2; 4 per cent); lithium, beryllium (element 4) and boron (element 5) were present only in traces. The list of chemical elements at this time was thus, effectively, only two members long, only one of which is chemically active (helium is chemically inert). The Big Bang brought matter into the Universe and started to mould it into the first few chemical elements, but it was only through its later consequences that the remaining elements came into being, giving us the total of ninety-four naturally occurring elements we have today, and others that are too short-lived to persist and were or are only temporary.

The Universe had to evolve for hundreds of millions of years before chemistry became interesting. There was hydrogen but certainly none of the other elements that make chemistry, including biochemistry, work. However, almost two-thirds of the number of atoms in the human body, and 10 per cent of its mass, are hydrogen atoms. Most of them were formed in the first few minutes of the Big Bang. Much of the material of our flesh is thus directly traceable back through the life of the Universe to its birth. Our chemical origins are, in part, literally in the Big Bang, although the Big Bang was not in itself enough to make all the chemical elements needed for us to exist.

The cosmic fireball and the Cosmic Microwave Background

In the cosmic fireball, all the newly created particles, both ordinary matter and dark matter, were rushing about in the Big Bang material. The dark matter WIMPs did not interact much, but the atomic particles from ordinary matter interacted with each other in frequent collisions, with the electrons absorbing and re-emitting photons. The photons were highly energetic light and X-rays. There is a similarity between the cosmic fireball and the intense flash of light and heat in a hydrogen bomb.

Unlike the brief flash of a nuclear weapon, however, the cosmic fireball lasted for about 380,000 years, the photons bottled up inside the cloud of electrons. But as time passed, the cloud expanded and became cooler and less dense. The electrons settled onto the hydrogen and helium nuclei and formed atoms. At that stage, the photons hitherto trapped inside the cloud by the free electrons were able to escape, the cloud having become transparent. Those photons were able to travel across space from then to now, a length of time of about 13.8 billion years. Travelling undisturbed, they thus preserved an imprint of what the Universe looked like when they became free and escaped and so carry a picture to us of the Universe as it was, aged 380,000 years.

There was one major change in the photons that did not disturb the image they carried but did alter the way we can detect them. They were degraded by the expansion of the Universe during their travel. The wavelengths of the photons were lengthened by the expansion, so that they became less energetic, changing progressively from X-rays to ultraviolet light to light to infrared radiation and ending up here and now as microwave radiation. This is the same kind of radiation that powers a microwave oven and is used to link the towers of telecommunication networks. Microwaves are radio waves that have a wavelength within the range of about 1 millimetre to about 1 metre. The microwave photons that originated in the Big Bang impinge on us from all directions in space and form what is known as the Cosmic Microwave Background radiation (CMB).

The CMB can be detected by radio telescopes and was first discovered in 1965 by two American radio engineers, Arno Penzias (b. 1933) and Robert Woodrow Wilson (b. 1936). They were systematically trying to identify all the sources of noise in a very sensitive receiver-antenna combination working at a wavelength of 7.3 centimetres. Working at the Bell Telephone Laboratories site in Holmdel, New Jersey, which was used for early experiments on communication satellites, they were able to eliminate or measure all the instrumental effects (including the effect of droppings from two pigeons that had decided to nest in the horn), but they always saw excess noise, which was not accounted for by any possible local source. They thought that it must have a natural origin and began to consider whether it was astronomical.

By chance, Penzias and Wilson stumbled across the origin of the microwaves when they were told by a colleague about remarks made by Princeton University physicist Jim Peebles (b. 1935) at an astronomy lecture at Johns Hopkins University in Baltimore. Peebles had described Gamow's ideas about the fireball of the Big Bang and how the radiation that it generated would look now. This set Penzias and Wilson to think about the cosmic fireball as the source of the natural noise that they had measured. They had made a double Nobel Prize-winning discovery. One award was given to Penzias and Wilson in 1978 for finding the radiation, one to Peebles in 2019 for explaining it even before it had been found.

The CMB is remarkably uniform. It comes almost equally from every direction and its temperature is almost the same everywhere. In fact, these are the main reasons why astronomers concluded that the CMB comes from the Universe as a whole. This uses a line of argument called the Copernican principle, reflecting the way that Polish cleric and astronomer Nicolaus Copernicus (1473–1543) proposed a change to the then accepted position of the Earth. His theory, published in 1543, displaced the Earth from the centre of the Universe, with everything in orbit around our planet, to a position as one of the planets that orbited around the Sun. A succession of scientific advances has confirmed our mediocrity ever since: we are not at the centre of the solar system, we are not at the centre of our Galaxy, we are not at the centre of the Local Supercluster of the galaxies that surround us. We are inside the Universe, but there is no particular significance to our viewing point. Seeing something that is almost completely uniform all around us could lead us to think that we are at its centre, but if we think that this is ruled out, as it is by the Copernican principle, it can only look uniform if it is uniform everywhere, the same throughout the entire Universe – the microwave background must be something that pertains to the cosmos as a whole; that is what led to the term 'Cosmic Microwave Background'.

Embryonic clusters of galaxies:
the first structures revealed

Although the CMB is extremely uniform, it is not *exactly* uniform. It could not be: the material of the Big Bang was composed of fundamental particles and radiation to which quantum mechanics applies, and quantum mechanics is probabilistic. There is no definite state in quantum mechanics for a material to be in, only a range of possible states, which are occupied each with a certain probability. So, one patch of the Big Bang material existed in one set of conditions (temperature, energy, density, pressure and so on) and another patch was different, although only slightly. As the Big Bang developed, the differences between patches intensified as the denser patches contracted under their own gravity. The minute fluctuations in the brightness, temperature, and so on, of the Big Bang material from one place amplified and grew into larger fluctuations. At a certain point the Big Bang material became so rarefied that it became transparent, letting out the radiation that became the CMB, and, as described above, this carried images of the Big Bang at the time it was released, including whatever structure was there at that time.

The size of the brighter and fainter patches and the level of non-uniformity were predicted theoretically by Russian astrophysicists Rashid Sunyaev (b. 1943) and Yakov Zel'dovich (1914–1987) in 1966–70. The prediction could not be very precise, but suggested that the level of the variation in brightness was about the same as the patchiness across the most uniform white paper – not very patchy.

Sunyaev is a large, affable bear of a man. Born in Tashkent, now the capital of Uzbekistan, he left in 1960 aged seventeen to study in Moscow, after promising his grandmother he would not work on bombs. Setting out on his PhD studies, he was told that astrophysics was useless and became intent on research into particle physics, but he was taken under Zel'dovich's wing in 1965 and was inspired by him to work on astrophysical problems; this redirected his career. That year, Penzias and Wilson discovered the CMB. Zel'dovich until then espoused a theory of the origin of the

Universe that started from dense material at a cold temperature. He immediately abandoned his own ideas and adopted the hot Big Bang model, starting to work out its properties. Together the two men worked to identify processes occurring in the Big Bang material and the way that small irregularities developed into large-scale structure.

Sunyaev eventually became the director of the Max Planck Institute for Astrophysics in Garching, near Munich. He is gifted with considerable powers of persuasion, and he used them effectively to inspire numerous teams of scientists to measure the irregularities that he expected in the CMB. There were many unsuccessful searches by telescopes based on the ground to discover these, targeted at various levels within the range of uncertainty, and an equally unsuccessful attempt by a Soviet scientific mission, Relikt-1, in 1983. These efforts narrowed down all the possibilities, so that there was only one range where the fluctuations could be: if they had not then been found, the Big Bang theory would have been in big trouble. The fact that fluctuations existed was established by NASA's Cosmic Background Explorer (COBE) satellite in 1989. The work was recognized by the award of the 2006 Nobel Prize in Physics to the leaders of the COBE satellite team, American astrophysicists John Mather, who coordinated the entire process and measured the properties of the radiation, and George Smoot, who measured its small variations.

COBE showed that the variations existed and measured them, but did not map them. The first map was made by NASA's follow-up mission to COBE, the Wilkinson Microwave Anisotropy Probe (WMAP; 2001–10). The Planck space satellite, which was launched by the European Space Agency (ESA) in 2009 and orbited until 2013, has provided the most accurate map (pl. IV). The measurements obtained by the satellite achieved the ultimate limit of accuracy at which the fluctuations can be measured (because of complicating interference from other astronomical radiations).

Maps of the CMB show graininess at the level of 1 part in 100,000. The graininess represents differences in density in the material of the Big Bang, originating from inevitable random fluctuations. The maps represent the Universe at an age of 380,000 years. As we shall see in the next chapter, the dense spots are embryonic clusters of galaxies.

The expanding Universe

In Chapter 1, the Big Bang was identified as the moment of birth of the Universe, but perhaps in some ways it would be better to call it its conception. The time of the cosmic fireball was the equivalent of its development as an embryo. Before the generation of the CMB it was a completely alien place of elementary particles and very energetic radiation. From 380,000 years old, it could be regarded as a child, beginning to function like a person, maturing and starting to become more familiar. The period starting from then has been one of progressive growth, in what will for the Universe be a never-ending expansion. Cosmology is the science of the Universe as a whole and attempts to describe the significant characteristics of that growth. It puts forward detailed, often speculative, descriptions of what the Universe is made of, like hydrogen gas, radiation and dark matter. It develops descriptions of how the Universe assembles its component materials into structures, like galaxies, and what then happens, such as the ways the structures interact. It also tries to fit everything into a moving picture showing how everything develops. This overall framework is called a model.

In past times the model would have been literally a model – a physical miniature toy made of wooden struts and metal gears, like a mechanical clock. In fact, in medieval times clocks were indeed actual models of the Universe as it was known then. They harked back to the way a sundial works, each made with a pointer rotating once in twenty-four hours and representing the motion of the Sun rotating around the Earth. Nowadays, a cosmological model is a set of mathematical equations that are given form as diagrams and graphs, but seldom if ever given a tangible form as a physical device.

Modern cosmology began in 1917 with Albert Einstein's paper about general relativity, 'Cosmological Considerations in the General Theory of Relativity' (see Chapter 1). Dutch mathematician Willem de Sitter, German physicist Karl Schwarzschild (see page 67) and Russian mathematician Alexander Friedmann developed cosmological models built on Einstein's foundation. It was found that general relativity admitted the idea of an expanding Universe, and this was almost immediately confirmed by the realization that galaxies were mileposts that delineated the great distances in space and were receding from our own Galaxy. In 1929, Edwin Hubble discovered that galaxies were receding in such a way that the speed of recession of a galaxy was proportional to its distance (see page 20). The constant of proportionality is called Hubble's constant.

Until 2013, astronomers used direct measurements of speeds and distances of galaxies to determine Hubble's constant. They can do this only for nearby galaxies by measuring the apparent brightness of Cepheid variable stars and supernovae, whose actual brightness is known somehow. In the same way, if you see a light at night, you can estimate its distance if you know that it originates in the dim flame of a match, the beam of a car headlight or the glare of a lighthouse. The technique was developed over the previous century, first for Cepheid variable stars, pioneered by American astronomer Henrietta Leavitt (1868–1921).

Leavitt was one of a large number of women scientists employed by Edward Pickering (1846–1919), the director of the Harvard College Observatory in Cambridge, Massachusetts. Williamina Fleming was another (see page 164). Leavitt, who was deaf, attended Radcliffe College where she was attracted to astronomy and spotted by Pickering. She came to head the department within Harvard College Observatory, which was carrying out a major photographic project organized by Pickering to measure the brightness of stars in the Milky Way and the two nearest galaxies – the Magellanic Clouds (see page 107). Using a technique called fly-spanking (looking through a microscope at the black spots left by the light from stars on photographic plates – they look like flies that have been swatted – and estimating their size, i.e., their brightness), Leavitt found and measured 1,777 variable stars on pictures of the Magellanic Clouds taken at the Harvard Southern Station in Arequipa, Peru.

The stars were variable stars of all classes, but some of them were similar to those in the Milky Way known as Cepheids (pronounced 'seff-ids'; Delta Cephei is the prototype star). Cepheids vary in brightness periodically in a recognizable regular cycle. In 1908, Leavitt plotted the average brightness of the Cepheid variable stars in the Large Magellanic Cloud versus their periods. There was a clear correlation. The period-luminosity (P-L) relationship was not detectable in Cepheids in our Milky Way because the stars were all at different distances as well as having differing intrinsic brightnesses, but the Cepheids in the Large Magellanic Cloud are all at the same distance. Leavitt's sample of Cepheids eliminated the distance variable in the plot and the P-L relationship revealed itself.

The P-L relationship became a way to estimate the intrinsic brightness of a Cepheid star – equivalent to some indicator about the origin of a light seen at night, such as observing whether it flickers (a flame), shines steadily (a headlight) or flashes periodically (a lighthouse). The method for this was to determine the period of variation of the star and use the P-L relationship to estimate its intrinsic brightness, then relate its intrinsic brightness to its apparent brightness and deduce the distance of the Cepheid and the galaxy that it inhabits (if it is intrinsically very bright and looks dim it must be far away, but if it is feeble and looks bright, it must be close). The discovery was the key to determining the distance to galaxies in terms of the distance to the Large Magellanic Cloud, and so a tool for establishing the distance scale of the Universe and therefore Hubble's constant.

Recent determinations of Hubble's constant have used the Hubble Space Telescope to redetermine the distance scale by the Cepheid technique. In fact, the telescope was launched in

1990 for this as its main purpose. Astronomers have also found parallel ways to tackle the same problem, including with the Gaia space mission (see pages 100–102). Cutting a longer story short, astronomers seem agreed that these techniques give a value for Hubble's constant of 73.5 kilometres per second per megaparsec, plus or minus 1.4 (a parsec is a measure of astronomical distance equivalent to 3.3 light years; a megaparsec is 1 million parsecs). So, if there are two galaxies, one of them 3.3 million light years further away from us than the other, then, because of the expansion of the Universe, the further galaxy is expected to move away from us at a speed that is 73.5 kilometres per second faster than the closer galaxy.

The expansion is the outcome of an explosion: the Big Bang. In an explosion in which various fragments are thrown out at different speeds, in a given time the faster bits have moved further than the slower ones. In fact, the distance travelled, divided by the speed of any fragment, is the time since the explosion. The significance of Hubble's constant is that it is related to the age of the Universe, with the value of 73.5 corresponding to an age of 12.7 billion years.

A completely independent way to determine Hubble's constant became available in 2009 when the Planck satellite was launched to investigate the CMB. It collected data until 2013 and it then took five years for a big team of astronomers to analyse the data to everyone's satisfaction. First off, astronomers had to decide exactly which cosmological model to use. Then they had to adjust the chosen model to fit the image and see how changing the value of Hubble's constant affects the outcome. This enabled them to bear down on the most appropriate model.

The model that worked the best, called Λ-CDM, was the one that had become the standard model even before Planck was launched, where Λ is the Greek capital letter lambda and CDM stands for 'cold dark matter'. Nearly all astronomers believe that dark matter exists but some believe in one kind of dark matter, others in another. Cold dark matter is made of fundamental particles that move at less than the speed of light, and has the favoured property that it helps produce galaxies and clusters of galaxies by growing them from merging smaller lumps. Λ is the cosmological constant that Einstein proposed in 1917 and imagined would help prop up the Universe and keep it static (see page 16). In modern cosmological models, it represents a similar push, but not one that keeps a static universe static. It is in effect a force that pushes on an already expanding Universe to make it expand faster.

The cosmological constant labels mathematically a feature of the expansion of the Universe that was discovered nearly a century later, in 1998–99. The Universe is accelerating. This is completely contrary to expectation. The galaxies and dark matter in the Universe mutually pull one another, a process that should slow the expansion down. When astronomers used the time machine that is the Hubble Space Telescope, they found the contrary. Gazing into the distant Universe, astronomers looked back in time, because light travels at a finite speed. The furthest galaxies, representing the earlier Universe, should be moving more quickly than nearby ones. In 1998–99, astronomers from the Supernova Cosmology Project and the High-Z Supernova Search Team used supernovae observed with the Hubble Space Telescope to check the distances of distant galaxies, coordinating with the largest ground-based optical telescopes to check how fast they were moving. The two teams discovered the reverse of what was expected and the three leaders, American astrophysicists Saul Perlmutter, Brian P. Schmidt and Adam G. Riess, were awarded the 2011 Nobel Prize in Physics for the discovery.

The expansion of the Universe is speeding up, not slowing down. There is some progressive input of energy into the expansion of the Universe. It goes under the name of 'dark energy' and its nature is a mystery; it is even more of a mystery than dark matter. In a Universe with both matter (whether dark or not) and dark energy, there is a competition between the tendency of Λ and dark energy to cause acceleration and the tendency of gravity and matter to cause deceleration. This has a big effect on the 'formation of structure', the term that astronomers give to the way that the earliest irregularities in the material of the Big Bang grew. The balance between gravitation and Λ controlled the way our Universe is developing.

The Millennium Simulation, a major calculation by astronomers at the University of Durham and the Max Planck Institute for Astrophysics (see also Chapter 3), showed how the formation of a cluster of galaxies starts weakly because the fluctuations in the Big Bang material are not very pronounced. Gravity, principally from dark matter, draws in surrounding material, but then the release of dark energy tends to stabilize the infall, which peters out as the galaxies orbit one another.

The characteristics of the picture of the CMB drawn from the Planck satellite data suggested that the Λ-CDM model is correct: the standard model of cosmology survived all the tests. Tongue in cheek, scientists praised the Universe because the real data fits their model so well and acclaimed the Universe as 'almost perfect'. The fit gave a value for Hubble's constant: namely 67.4 kilometres per second per megaparsec, plus or minus 0.4, which corresponds to an age for the Universe of 13.8 billion years. The Planck result for Hubble's constant is near to the value from measurements using the Hubble Space Telescope and was a cause for early celebration. A closer look is more sobering, however; in fact, the Planck value is worryingly different from the Hubble Space Telescope's.

On the one hand, it is fantastic that two completely different ways of tackling the problem – one focusing on the nearby, old Universe, the other on the distant, young Universe – are close to each other. On the other hand, the two figures do not agree to within their respective uncertainties and that constitutes a discrepancy to be resolved. The discrepancy might be due to some problem of a technical nature. Or, it might mean that there is something fundamentally wrong with the science behind the Λ-CDM model – we may need 'new physics'. The problem has not been settled.

The Planck data provides values for other parameters in the cosmological models that are interesting answers to basic questions, for example:

- Will the Universe expand forever or will it eventually coast to a halt and collapse? This depends on the density of the Universe, and the Planck data suggests that the Universe is just on the borderline and will slow its expansion but never completely stop (this is explored further in Chapter 12).
- How much dark matter is there in the Universe in proportion to ordinary matter? There is five and a half times as much.
- How much matter, dark matter and dark energy are there in the Universe? Ordinary matter makes up 4.9 per cent of the energy in the Universe, dark matter 26.8 per cent and dark energy 68.3 per cent.

Planck has initiated the present scientific era in which cosmology has become an accurate science rather than speculative theories. Cosmologists have become more confident about their subject. However, we are disappointingly ignorant about the content of 95 per cent of our Universe. At least we know how ignorant we are.

3
Randomness Becomes Structure: The Formation of the First Galaxies

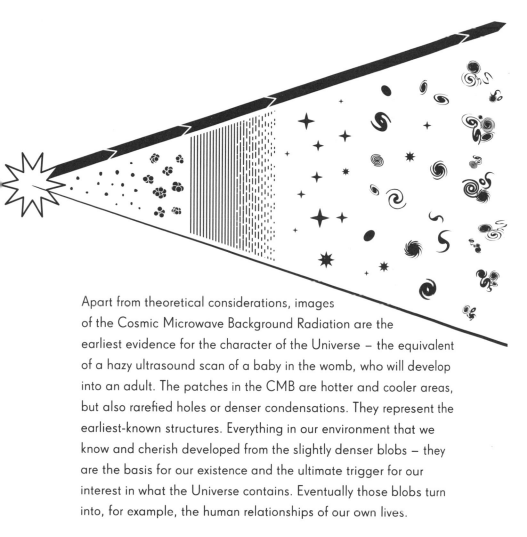

Apart from theoretical considerations, images of the Cosmic Microwave Background Radiation are the earliest evidence for the character of the Universe – the equivalent of a hazy ultrasound scan of a baby in the womb, who will develop into an adult. The patches in the CMB are hotter and cooler areas, but also rarefied holes or denser condensations. They represent the earliest-known structures. Everything in our environment that we know and cherish developed from the slightly denser blobs – they are the basis for our existence and the ultimate trigger for our interest in what the Universe contains. Eventually those blobs turn into, for example, the human relationships of our own lives.

The Big Bang created matter, dark matter and radiation. The radiation faded away, and stars formed in small galaxies, which merged to make larger ones. Galaxies then congregated into groups and clusters of galaxies.

Randomness and pattern in the early universe

What happened all those billions of years ago was that a slight excess in the density of a blob, an excess both of ordinary matter and dark matter, caused an inward, attractive gravitational force on the material surrounding the blob. This drew the extra material inwards and caused the blob to condense further. Thus, the blob grew in both mass and density. The development of the blobs evolved into a web of lumpy filaments connecting the larger blobs.

Over the first 200 million or 300 million years of the life of the Universe, the blobs and the filaments went on to form concentrated masses of matter – intergalactic clouds of dark matter and a mixture of hydrogen and helium. The matter and the dark matter cooperated, the one helping the other: the dark matter dragged in the matter and the matter augmented the inward gravitational attraction of the dark matter. The matter also resisted the gravitational force with an outward pressure. The most massive blobs were to form the largest structures in the Universe: clusters of galaxies. Within the clusters, further concentrations formed galaxies, within which developed stars and planetary systems – and ultimately ourselves. Galaxies also formed mostly in the filaments.

Although the blobbiness in the CMB is random, in the sense that there was apparently no regular pattern, it does have some discernible general characteristics: the blobs have a typical size and separation. This is the result of a marvellous effect called acoustic oscillations: the early Universe rang with sound! It did not play just a single note, such as a flute would make, but a range of notes, somewhat similar to the sound that would be created by a trickle of sand falling onto a drum. The notes have persisted to this time as a faint whisper discernible in the arrangement of galaxies, just as the sand would collect preferentially in various places on the drumskin.

The pattern of notes has been brought out in a statistical analysis of the CMB called cross-correlation. Imagine a simplified pattern, in which the CMB is made of bright and dark squares, like a chessboard, with the white representing the bright areas and the

I The Milky Way. Star clouds towards the centre of our Galaxy arc across the dawn sky at the La Silla Observatory, in Chile. Dark holes in the massed stars, in reality clouds of dust particles, block the light behind them. The triangular white glow that stretches up and to the left from the orange glow that marks where the Sun is, below the horizon, is known as the zodiacal light. It is dust in the solar system, ground from the surface of colliding asteroids and comets. The 15-metre Swedish–ESO Submillimetre Telescope stands in the foreground, apparently pointing to the planet Venus.

II The Hubble Ultra Deep Field. The Hubble Space Telescope targeted a small patch of what was up to then empty sky, and imaged it over and over again in a deep exposure to reveal 10,000 of the faintest, furthest galaxies. The group photograph tells the biography of galaxies as they evolve over 13.2 billion years of cosmic time – newborn infants, adolescents and adults – and confirms the black sky between the galaxies that is the clue that the Universe is not infinitely old.

III Abell 1689, one of the biggest clusters of galaxies known, 2.4 billion light years away. The photograph shows not only the galaxies in the cluster but also a system of streaks and arcs that are the distorted images of galaxies lying far in the background and magnified by the gravitational lens effect of all the matter in the cluster. The purple overlay represents the matter, both ordinary and dark matter.

IV Cosmic Microwave Background. The most detailed map of the CMB made by the Planck satellite shows warmer and cooler patches (red and blue), which represent clumpiness of the Big Bang material some 380,000 years after the initial explosion.

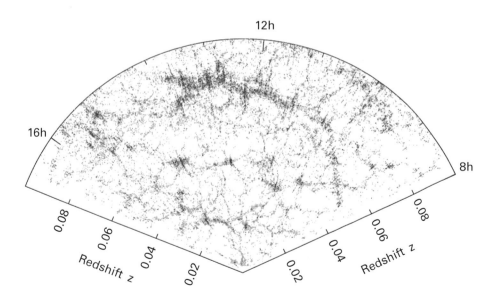

V The cosmic web. A map of the galaxies in a thin wedge of the sky, centred on the Earth and extending to 1.3 billion light years away. Galaxies form clumps, filaments, and sheets, interweaved with voids in the shapes of bubbles and tunnels.

VI Antenna Galaxies. NGC 4038 and NGC 4039 were spiral galaxies, but a few hundred million years ago they began to collide violently, throwing stars out into intergalactic space and each triggering a burst in each other of bright blue stars that excite clouds of gas into bright pink and red colours. Eventually, order will be restored and the two galaxies will have transformed into one large elliptical galaxy.

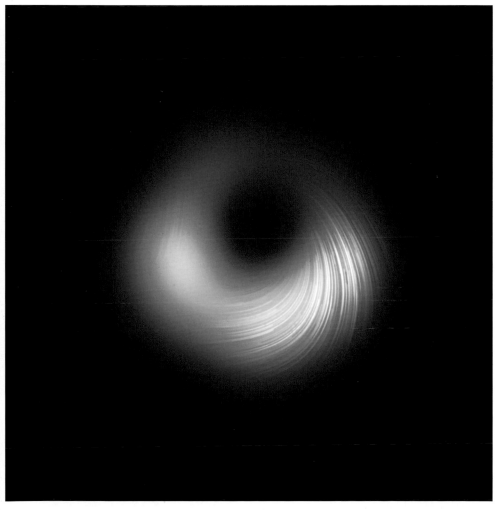

VII Black hole. A disc of hot material swirls around the supermassive black hole in the galaxy M 87. The literal black hole in the centre of the disc is a volume devoid of material (which has all fallen on to the black hole) and, at its very centre, the physical black hole enclosing a zone inside the event horizon, from which no radiation escapes. Spiralling streaks map the magnetic field created by the black hole.

VIII M 83. This spiral galaxy is thought to mimic our own. It has two main spiral arms emanating from the ends of a central bar of stars. Dark dust clouds and pink nebulae delineate the inner edges of the spiral arms.

IX Pillars of Creation. Columns of gas and dust protrude from the Eagle Nebula. Newly formed stars hide in their tips. The gas and dust in the pillars is scoured by intense radiation and winds from massive, young stars nearby, eroding the surface of the pillars into wispy tendrils of cosmic dust and elephants' trunks.

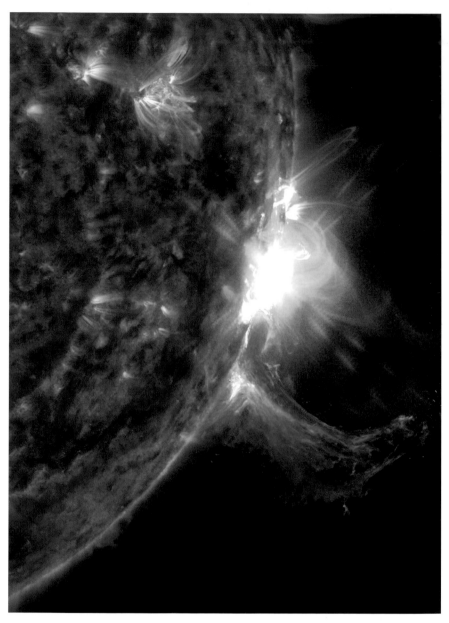

X Solar flare. The Sun emitted a mid-level solar flare on 2 October 2014, recorded as a bright burst by NASA's Solar Dynamics Observatory. The same paroxysm of the solar surface ejected solar material into interplanetary space in a prominence, which can be seen below the flare in this image.

black the dark areas. Choose two points that are a certain distance apart and multiply their brightness together: if one point is in a white square (call its brightness 1) and the other is also in a white square, the cross-correlation is $1 \times 1 = 1$. All the pairs of points that lie within the white squares will produce the same result. But if the second point falls within a black square (brightness 0), the cross-correlation is $1 \times 0 = 0$. Repeat this with all the points that are at the same separation, then add up all the cross-correlation products. Now do it all again starting with a different point. When you have done this with all the possible starting points, choose a different separation and repeat the process. Now plot the cross-correlation as a function of separation. For a chequer pattern, there will be a peak that corresponds to separations that are up to the size of the chequers, then the cross-correlation will dip when the separations stretch across one chequer into squares of the opposite colour.

The cross-correlation pattern of the CMB shows that the most frequent size of a blob is 1 degree. Why is this? The reason is that gravitation was not the only force influencing how the blobs grew in size. Pressure forces also had a role. The light and heat trapped in the moving plasma of the early Universe as a blob became denser caused an increase in pressure, which generated a pressure wave – in physics, a pressure wave in material is called a sound wave. The matter was driven by the radiation and tried to expand, sending the sound wave into the surrounding material of very dense plasma – a sound wave that travelled initially at one-third of the speed of light! However, dark matter does not interact with radiation – so although the dark matter responded to changes in gravitational force caused by changes in the density of the matter, it ignored the radiation. Therefore, the denser blobs of matter attracted both ordinary matter and dark matter, but not in the same way.

After about 380,000 years the electrons and protons combined. The matter in the Universe had been a plasma, but it became an electrically neutral gas of atoms, which did not interact as much with the radiation, so the sound wave stalled. This froze the regular structure of the waves in the gaseous material in a pattern of what are known as acoustic oscillations. The series of the many waves overlapped, contributing to a random pattern. Earlier, this was likened to the effect of sand falling on a drumskin, but this analogy was deficient in that the drumskin is stretched on a fixed, circular framework and the pattern of oscillations is regular – it is this property that enables a musical instrument to produce a pure musical note. There is no such fixed boundary in the Universe, and a better analogy might be to think about the oscillations on the surface of a large pond. If one region produces an effect like throwing a pebble into the pond and seeing the expanding ripple, then the Universe as a whole produces the effect of throwing a handful of gravel into the water and looking at the overlapping choppy disturbances.

The Universe is thus, in this way, random, but nevertheless, statistical traces of the simple patterns of which it is composed remain – within the noise of the random structure there are even now the traces of a vestigial note. There is a likelihood that if there is a dense spot in the CMB, other dense spots lie within a 500-million-light-year radius. The blobs of Big Bang material became patches in the CMB, and the 500-million-light-year dimension corresponds to the angular size of 1 degree, which is typical of the size of a patch.

When the blobs of the CMB condensed into galaxies a few hundred million years after the CMB became free of interaction with the material of the Big Bang, the galaxies preserved the tendency to be clumpy on this scale and the clumpiness persists even today. If a galaxy is discovered, there will be galaxies all around, distributed at random, but there is a slightly larger than random statistical chance that there is a galaxy within 500 million light years. The apparently random distribution both of the patches in the CMB and of galaxies shares the same underlying pattern of clustering.

The clustering of galaxies was first hinted at in an analysis by British cosmologists John Peacock and Shaun Cole in 2001, and then clearly detected in 2005 by a team led by American cosmologist Daniel Eisenstein of the Steward Observatory, University of Arizona. They analysed the positions in three-dimensional space of

46,748 galaxies distributed over 9 per cent of the sky, in a volume equal to a cube measuring 4 billion light years on each side. Because of the capacity of modern computers, it is now common for large teams to acquire and analyse 'Big Data' from extensive surveys to find out things about the Universe – although it seems likely that there will always be a place for the lone scientist with the striking and original idea.

The structures of clusters of galaxies, superclusters and voids

The combination of randomness and pattern shows itself in the present distribution of galaxies: they group into strings and clusters. Clusters of galaxies are the largest structures in the Universe that are bound together by their mutual gravitational attraction as long-lasting entities. Even larger structures exist. They are known as superclusters – clusters of clusters of galaxies. They are in the process of dissipating, because their self-gravity is not strong enough to hold the component clusters of galaxies together. They may be the biggest structures in the Universe, but they are transient.

Within a supercluster, clusters of galaxies are connected by threads of galaxies, like the strands of webs. Sometimes the threads are flat and akin to ribbons, like the places where the surfaces of bubbles touch in a soapy foam. The clusters and superclusters are the regions where the surfaces of bubbles touch. Conversely, there are large volumes of space, like the interior of bubbles in the foam, in which galaxies are few and far between: these are called voids. All this structure developed from the random blobbiness in the fireball of the Big Bang, which imprinted the patches in the CMB. Astronomers have mapped and investigated some of the individual structures in the distribution of the galaxies that surround us. Some of the nearer structures of this cosmic web have been identified by plotting galaxies on a map (pl. v).

The first attempts to map the large-scale structure in the distribution of galaxies around us were led by American astrophysicist Margaret Geller (b. 1947) and astronomer John Huchra (1948–2010) of the Harvard & Smithsonian Center for Astrophysics (CfA) in the 1980s. The CfA survey probed out in a slice of the sky to 500 million light years and in about 1989 what became known as the Great Wall of galaxies, at a distance of about 400 million light years, was discovered. This was a line of galaxies roughly circumferential around us, and stretched from one edge of the survey to the other, so approximately 400 million light years long. Although named a 'wall', it was not very high; it could be imagined as a low dry-stone wall running across moorland, the kind that keeps sheep from straying too far. It is shaped somewhere between a wall and a filament.

The CfA survey was inspirational. In its second phase, it went on to accumulate the distances of 18,000 galaxies, and analysis showed that surveys of hundreds of thousands of galaxies or more were needed to make a good attempt at mapping local structure. This required mass-production techniques to be applied to a large fraction of the sky. The 2dF Galaxy Redshift Survey set out to do this. It was carried out between 1996 and 2001 in New South Wales by the Anglo-Australian Observatory with an amazing machine built by astronomer Keith Taylor that was carried at the prime focus of the Anglo-Australian Telescope. It used a robot to place optical fibres in the focal plane of the telescope to pick up light from four hundred galaxies at once, measuring redshifts at the rate of thousands per night. The redshift of a galaxy is the amount by which its spectrum is displaced from its normal position by virtue of its motion, mostly because the Universe is expanding. The key fact that makes a scale map of the Universe possible is that, because of the expansion of the Universe, the bigger the speed, the more distant the galaxy. Taylor's robotic machine was a device for showing how galaxies are distributed around us. (See Glossary for a further explanation of redshift.)

In a few hundred nights of observing spread over five years, the telescope measured 221,283 galaxies. Brought to fruition by an Australian-British team led by astronomer Matthew Colless of the Australian National University, it was the first galaxy survey to give a comprehensive and representational map of a significant fraction of the Universe. It not only mapped out the local voids, superclusters and filamentary walls that surround us, but also showed how the largest structures were compressed as the galaxies fell together, continuing the process of gravitational attraction that set off from the density fluctuations in the material of the Big Bang.

Even this survey was not enough to satisfy astronomers' needs. Spurred by the difficulties faced in 1990 by the Hubble Space Telescope when it was failing to deliver sharp images because of a manufacturing fault, Princeton University astronomer Jim Gunn (b. 1938; see also page 83) set up an independent project to locate and measure galaxies by the million. It was funded in major part by the Alfred Sloan Foundation and became known as the Sloan Digital Sky Survey (SDSS). The survey used a dedicated telescope and camera built for the purpose at Apache Point Observatory in the Sacramento Mountains of New Mexico. However, in contrast to a massive project like NASA's Hubble Space Telescope, to avoid stifling and diluting the scientific focus, the SDSS was deliberately kept to a scale and with a purpose that could be controlled by an individual scientist.

The SDSS telescope is not of world-beating size. Its mirror is 2.5 metres (100 inches) in diameter, ranking perhaps number 50 in the world. It is able to see a large area of the sky in its field of view, which is desirable in a telescope that is going to try to record everything. The camera is novel and uses huge CCDs (charge-coupled devices) to record the brightness and spectrum of what it sees in the sky. It is operated a little like Google Street View, which records places along streets driven by a mobile camera. The SDSS observing technique is to cause the telescope to scan at a precisely controlled rate along a track in the sky and to shift the image on the CCDs electronically at an exactly compensating rate to build up an exposure. In its first five years of operation, the project recorded the brightness of 1 billion stars and galaxies, and the spectrum of 4 million, a number that is increasing as the project continues.

The data from the SDSS is accumulated in a publicly accessible archive as soon as it is obtained and processed. It is unusual for data from an astronomical project to be made public so quickly – usually the project personnel have rights to withhold the data for a period of time as a scientific reward for putting in the effort to bring the project to fruition. The logic was that the project was funded by public sources of money and its data should be publicly available, and that it would be best for science if anyone could bring ideas to the archive to investigate its scientific possibilities. The project personnel knew so much about the instrument's capabilities and its programme, it was argued, that they had an advantage over the rest of the community of astronomers and ought to be able to make killer discoveries even if they were competing in the same time frame as everyone else. It has been a community effort to master the data produced by the SDSS and use it so successfully to map the structure of the Universe as it is known today.

An early success for the project was the identification of the Sloan Great Wall in 2003, a filament of the cosmic web more than 1 billion light years long and curved in an arc around us at a distance of 1 billion light years. Perhaps the most important discovery so far is the use of the data by Daniel Eisenstein to nail down the residual effects of acoustic oscillations in the distribution of galaxies. Eisenstein had shown how well attuned he was to the aims and operation of the SDSS project and became its director from 2006 to 2010.

The real Universe and the Millennium Simulation
Maps like the ones produced from the SDSS that showed how galaxies are distributed into characteristic shapes constitute a challenge to astronomers to explain how this came about. They were able to do so by using simulations in computers. The Millennium Simulation was a calculation run by the Virgo Consortium, centred on the Institute for Computational Cosmology at the University of Durham in the UK and the Max Planck Institute for Astrophysics in Germany, which showed how the small irregularities developed in the matter exploding in the Big Bang. The simulation took the

following form: create in a computer a virtual box a few billion cubic light years in volume, then scatter into it 10 billion particles representing clumps of matter and dark matter; programme the computer with physical laws such as gravity and watch how the particles pull each other, move and develop structures.

The resultant distribution of particles looks like a web when displayed as a two-dimensional cross-section or projection. In a three-dimensional representation it would look like a foam, with holes that now have a characteristic size of 500 million light years. The web of galaxies is not a rigid structure and is flexing and rippling with the residual random motions of the Big Bang and the subsequent cumulative effects of the pull of each galaxy on the rest.

Thus, these structures are transitory. They are large and the force of gravity that holds them together is weak. Moreover, the galaxies within them are moving quickly so they tend to dissipate. They have not dissipated yet because not enough time has elapsed since the time that they formed, just a few hundred million years after the Big Bang. However, within the structures are more compact groups of galaxies that will endure: they are the densest clusters of galaxies and are comparatively tightly packed together.

Dark matter in the cosmic web

Both the Millennium Simulation and the measurements of the CMB reveal the effects of dark matter on the cosmic web. The dark matter behaves differently from matter because, although both make their presence known by their gravity, they behave differently in their interaction with radiation and heat energy. Dark matter does not interact with light – which is why it is dark – but nor does it interact with infrared radiation or heat, so it does not cool as quickly as ordinary matter. Ordinary matter in the Big Bang material cools and gets denser and the dark matter remains hot and more diffuse, while both respond similarly to gravity. As a result, ordinary matter and dark matter concentrate into blobs in slightly different ways.

The differences were evident in the output of the Millennium Simulation. The dark matter blobs are in general bigger and more diffuse. Concentrations of ordinary matter like galaxies or clusters of galaxies are embedded in larger clouds or haloes of dark matter. This shows in the real Universe through gravitational lensing, which reveals where both types of matter are distributed in galaxies (pl. III). The dispersal of starlight shows where ordinary matter is located. In this way, astronomers can distinguish ordinary matter and dark matter and map the separate distribution of each kind.

Gravitational lensing is a phenomenon in which gravitation affects the path taken by light, deflecting it from a straight-line path. We ordinarily visualize space as 'nothing', measured by a kind of invisible rigid scaffold of three dimensions on which matter moves about like building workers. Albert Einstein had a more integrated view: space interacts with matter, each affecting the other. 'Matter tells space how to curve, space tells matter how to move,' wrote American physicist John Wheeler in 1973, explaining how Einstein's general theory of relativity works. In this way, Einstein offered a solution to one of the dilemmas of Newton's theory of gravity – how the force of gravity is transmitted through space.

Space is not 'nothing', it is a physical entity. Ordinary matter and dark matter alike bend space; the curvature of space bends the trajectories along which matter moves. Indeed, matter bends the trajectories of rays of light, just as glass does when shaped in a lens. Because matter is not uniformly distributed in the cosmic web, it bends rays of light from distant galaxies with a cumulative distorting effect as the rays pass through the Universe. We do not view distant galaxies, or indeed the image of the blobs of the CMB, through the perfect lens of nothing at all. We look back into the past life of the Universe through a crinkly lens of randomly distorted space, as if through a pane of bathroom glass. By analogy with optical lenses, this phenomenon is called gravitational lensing.

Just as a bathroom window obscures what lies beyond it, breaking up and rearranging the image seen through the glass, gravitational lensing limits the precision with which we can see images of distant cosmic history. At the same time, it provides a unique chance to study the distribution of matter in the cosmic web, giving us a way to map that matter. In principle (although I have never heard of anyone, even the most intrusive voyeur, actually doing this), one could learn about each tilted facet of the bathroom window pane by examining the fragmented image seen through it, straightening each fragment and reassembling the image clearly. In the astronomical case there are simplifying assumptions to be made about the pattern of lenses that are valid enough to calculate how each facet of the gravitational lens might affect the image, both to reassemble the distant image and to learn about the matter that caused the distortion. And because ordinary matter and dark matter both contribute to the distortion, and there is more dark matter than ordinary matter, it is a way to map the invisible dark matter that fills the otherwise vacant spaces of the Universe.

The first results of programmes intended to make such maps began to appear in the 1990s and pointed the way for later investigations. ESA's Planck satellite imaged the CMB (see Chapter 2) and in 2015, scientists analysing the data determined how the dark matter generally distributed in the Universe distorted the blobs in the image.

Other surveys doing much the same thing have used ground-based telescopes. One such, located at the European Southern Observatory's Paranal Observatory in Chile, is called the Kilo-Degree Survey (KiDS) and has imaged the galaxies in an area of the Southern sky about 1,500 square degrees in size (about 3 per cent of the entire sky). The light from the most distant galaxies (not as far away as the CMB, but far away nonetheless) is altered by gravitational lensing. Generally speaking, the distortion is very slight, but there are some dramatic individual cases, known as strong gravitational lensing. The Dark Energy Survey is similar to KiDS. It is based at the Cerro Tololo Inter-American Observatory, also in Chile, and has mapped an even bigger area of 5,000 square degrees for an equivalent analysis.

Both surveys come under the heading of 'Big Data'. Their analysis is highly technical and time-consuming because the data sets are large and complicated. The first images produced by KiDS – made by analysing the light collected from over 3 million distant galaxies more than 6 billion light years away – show concentrations of matter and dark matter in the blobs and filaments that are predicted. The images are broadly similar to the Millennium Simulation.

Having honed their techniques on ground-based telescopes, astronomers are planning to repeat the measurements in space. A new ESA cosmology satellite called Euclid is scheduled to launch in 2022 to measure this effect. The aim eventually is to do more than refine our knowledge about the way that matter in the Universe is distributed. It is to look at the way the cosmic web changes with time, developing over the life of the Universe, from the CMB to now. This history is important because the gravitational attraction that dark matter provided made the development of structure happen faster and stabilized the structure that developed.

Without dark matter playing its part in making galaxies and holding them together, we would not be here. Even though we do not know what dark matter is, and even though it now plays no significant part in our lives as we live them in our immediate environment, we owe our existence to its historic effects during the first 1 per cent of the expansion of the Universe.

The Virgo Cluster
The Virgo Cluster is our closest cluster of galaxies. It is at a distance of approximately 60 million light years from Earth spread over the constellations of Virgo and its neighbour Coma Berenices. The Virgo Cluster was first remarked on in 1784, when French astronomer Charles Messier (1730–1817) was compiling a catalogue of 'nebulae'. Messier was an assiduous comet hunter, scanning the sky night after night to discover new ones. When first seen, on the edge of the solar system far from the Sun, comets have no tails and are small, fuzzy objects looking very similar to some other celestial patches of light, which, in contradistinction to point-like

stars, were called *nebulae* (the Latin word for 'clouds'). If he came across such an object, Messier needed to spend time finding out whether it was a comet or a nebula. He compiled a catalogue of his own discoveries and those of his colleagues so that the nebulae could not be confused with comets in future.

Messier noticed that an unusual number of nebulae in the catalogue were grouped together in Virgo. What Messier called nebulae were, in fact, a mixture of true nebulae (clouds of interstellar gas or dust), star clusters and galaxies, which were all confused together in Messier's time. The nebulae in Virgo that formed the group identified by Messier are all galaxies. He made his remark in the catalogue's ninety-first entry, M 91: 'The constellation Virgo and especially the northern wing is one of the constellations which encloses the most nebulae.... All these nebulae...can be seen only in a good sky and near meridian passage. Most of these nebulae have been pointed out to me by M[onsieur] Méchain.'

Pierre Méchain (1744–1804) was a friend and colleague of Messier's. In a letter of 1783, published in the *Memoirs of the Berlin Royal Academy of Sciences and Arts*, he says that he had discovered more nebulae in this region: 'M. Messier mentions several nebulae in Virgo which I have indicated to him; but there are others which he has not seen.' No records of Méchain's additional observations are known.

In his catalogue, Messier listed a total of 109 objects of which 16 are, in modern terminology, galaxies in the Virgo Cluster and not nebulae at all. They are number 49 in the catalogue (M 49), followed by M 58, M 59, M 60, M 61, M 84, M 86, M 87, M 89 and M 90, which all lie in the Virgo constellation and M 85, M 88, M 91, M 98, M 99 and M 100, which lie within the present boundaries of the constellation of Coma Berenices. Messier numbered the celestial objects in his catalogue as he entered them, so M 49 was the first Virgo Cluster galaxy, discovered by Messier himself. He went on to discover M 58, M 59, M 60 and M 61 while following the track of the comet of 1779. German astronomer Johann Koehler and Italian Barnabus Oriani were doing the same thing and independently discovered M 59, M 60 and M 61. M 85 was discovered by Méchain and confirmed by Messier a few days later, together with M 84 to M 91, on the same night in March 1781. M 98 to 100 were also discovered by Méchain on one night and included by Messier as the last objects in the third edition of his catalogue just before he sent it to press. The most famous and most prominent Virgo Cluster galaxy is M 87, which is a giant elliptical galaxy that emits copious amounts of radio waves and is thus also known as the celestial radio source Virgo A.

It was in the first two decades of the twentieth century that these 'nebulae' revealed their true nature as galaxies, or star systems. Photography not only made possible this discovery by providing a means to measure the brightness of stars in galaxies and thus show how distant they are, but it also enabled the discovery of fainter galaxies. The number of galaxies increased from the score or so that could be seen by Messier and Méchain by eye to the hundreds. The first investigations of the nature of the Virgo Cluster were reported in an impressive series of papers by the Harvard College Observatory astronomers Harlow Shapley (1885–1972) and Adelaide Ames (1900–1932), the data derived from one- to two-hour exposures of photographs taken with various telescopes at Arequipa in Peru. In 1926, Shapley and Ames counted 103 spiral galaxies in the cluster, within an area of 100 square degrees, noting that there were many more too faint to see clearly.

Shapley led the work. He had started his career as a journalist and intended to develop his credentials in the new School of Journalism at the University of Missouri but, finding that the new school was not yet open, he turned in 1908 to astronomy. In 1920, he participated with Henry Curtis in what became known as the Great Debate at the National Academy of Sciences in Washington DC on the scale of the Universe, which established that the Milky Way was a galaxy of stars and that many 'nebulae' were similar galaxies beyond the Milky Way – so-called 'island universes' (see page 98). His performance in this debate was impressive enough that he was appointed as director of the Harvard College Observatory.

He became a very influential figure in American astronomy, and science in general, with his fingers in many pies, taking on many projects and administrative tasks.

In 1923, Shapley hired Ames as a research assistant: she, too, had intended to have a career as a journalist but, upon graduation from Vassar College in New York State, was unable to find work in journalism. Sadly, her career and her life were cut short prematurely in 1932 when, on a lakeside holiday, her canoe overturned and she drowned, aged thirty-two. The job offered to Ames was to establish a list of the known galaxies and measure their properties. The year before her death, Shapley and Ames had published a catalogue of 2,800 galaxies, of which 1,246 near the north pole of the Milky Way formed a more detailed list in what has become known as the Shapley-Ames Survey. It is still used today. Ames's background as a journalist formed her attitude to her astronomical work. 'I collect only the facts,' she is recorded as saying. 'The theories are Dr Shapley's.' She took pleasure in her astronomical work: 'For constant thrill I'd hand it to journalism, but for lasting satisfaction, give me astronomy!'

Modern surveys identify up to two thousand galaxies in the Virgo Cluster (because it is the nearest cluster of considerable size, it has been very well studied). It consists not only of galaxies – between the galaxies are many intergalactic stars (perhaps these constitute as much as 10 per cent of the number of stars in the galaxies) and a hot, rarefied plasma of intergalactic gas. The stars have been thrown out of their parent galaxies as they jostled past other galaxies. The plasma has a temperature of 30 million degrees Celsius (54 million degrees Fahrenheit). It emits X-rays and is heated by the black hole in the core of M 87 (pl. VII) and the passage of other galaxies through the cluster that cause the gas to slosh back and forth.

The Virgo Cluster is a part of the Local Supercluster of galaxies, also known as the Laniakea Supercluster, the name deriving from two Hawaiian words, *lani*, meaning 'sky, heaven', and *akea*, meaning 'broad, wide, spacious, immeasurable'. It was mapped by a University of Hawaii team led by Canadian astronomer Brent

Tully and contains hundreds of thousands of galaxies. Our own Galaxy is situated on the boundary between this supercluster and a large empty space called the Local Void.

The Great Attractor

Our Galaxy and the others nearby feel the gravitational pull of the galaxies all around. That pull is unbalanced because there is an uneven distribution of galaxies around us – more on one side than the other. As a result, our Galaxy is travelling through space at a speed of 600 kilometres per second (1.3 million miles per hour), additional to the motion it has as part of the expansion of the Universe. This streaming motion was identified in 1987 by the 'Seven Samurai' – not the group of Japanese warriors in the 1956 movie by Akira Kurosawa, but a group of astronomers, namely David Burstein, Roger Davies, Alan Dressler, Sandra Faber, Donald Lynden-Bell, Roberto Terlevich and Gary A. Wegner.

The direction towards which our Galaxy is streaming lies behind the galactic centre and is both confused and obscured by stars and dust in the Milky Way. Additionally, the surveys of galaxies at the time were less extensive than they became. The Seven Samurai were clear in 1987 that the streaming motion is real and is connected with the large number of galaxies in the direction towards which the stream is heading (a dense scattering called the Shapley Concentration, a term originating from the work of Shapley in 1932). However, they could not identify precisely what causes the motion. During a press conference after a lecture about the issue at the American Astronomical Society, American astronomer Dressler, impromptu, referred to it as an unknown entity using the name 'the Great Attractor' – some enormous thing in the Universe that is tugging on our Galaxy. If he had used 'Cluster X', or even 'the Shapley Concentration', it would not have caught the imagination, but the Great Attractor was too spectacular a concept for the press to ignore and sparked articles about its role in the astrological charts of potential life-partners or in the final destiny of the solar system. The name had slipped out from Dressler's mouth unpremeditated and his colleagues in the Seven Samurai expressed concern that the flashy name might detract from the scientific content of the work, although they also recognized that it would ensure that the work would become widely known. 'My fellow Samurai were happy to pass on both the credit and the blame,' wrote Dressler, wryly.

It took thirty years to map out specifically what is happening. The Great Attractor is not a monster black hole, or anything similar. According to Tully and his collaborators, the motion of the Galaxy is the result of a combination of the effects of the pull of the Virgo Cluster of galaxies and the Laniakea Supercluster, whose central concentration is the Great Attractor, coupled with the emptiness of the Local Void and the lack of compensating pull from galaxies in that direction.

One general conclusion that can be inferred from this particular instance is that galaxies are still being drawn towards concentrations of other galaxies. In other words, the voids of the cosmic web are growing: space is getting emptier.

4

The Dark Ages and the Emerging Cosmic Dawn

Between the emergence of the Cosmic Microwave Background Radiation out of the maelstrom of the Big Bang and the formation of the first galaxies came a period known to astronomers as the Dark Ages, when no light is visible from galaxies or quasars because there were fewer or no galaxies or quasars. Even when there were, they were shrouded and hidden in obscuring dust. This chapter is the story of why the Dark Ages were dark, and how the darkness cleared.

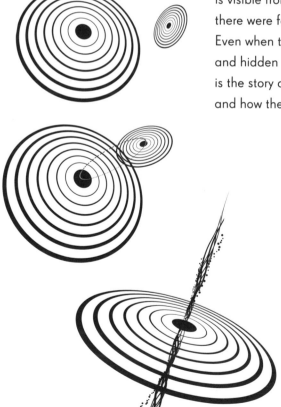

Black holes, surrounded by orbiting discs, merged to become powerful, radiating quasars.

The formation of the first stars

It took perhaps a few hundred million years to make the first stars and galaxies that shone their light out into space – an epoch termed 'Cosmic Dawn'. Even when stars appeared, however, their light did not penetrate very far into space because it was absorbed by surrounding gas and dust.

The first galaxies were made solely of hydrogen and helium, the only elements made in the Big Bang, so, naturally, their stars were made just of these elements too. One consequence of this chemical purity is that the stars in these galaxies had no planets, certainly not planets like the Earth. Terrestrial planets are made of heavier elements than hydrogen – for example, as we will see later in Chapter 10, the Earth has an iron core: no iron, no Earth. Moreover, all planets have been built up at the start of the planet-building process by sticking solid particles of heavy elements together (see Chapter 9). So, it is likely that a star that is made from pure hydrogen and helium will have no planets whatsoever, not even a jupiter, even though jupiters are predominantly hydrogen and helium.

Another feature of the first stars is that they were more massive and brighter than the stars that are typical now. The temperature and pressure inside a massive star are much higher than inside a less massive one. The nuclear fusion reactor in each star's core is thus more active, generating more energy. As the energy flows up through the body of the star it pushes upwards due to the phenomenon of radiation pressure. This supports the body of the star against the downward force of gravity and holds the star up, stopping it from collapsing. But if the star is too massive, it generates too much energy. Radiation pressure overcomes the star's gravitational force and blows the star apart. There is thus an upper limit to the mass of a star.

The exact value of the upper limit to the mass of a star depends on its chemical composition, because atoms of different structures intercept more or less radiation and this changes its upward pressure. The Sun is made of hydrogen and helium with just 2 per cent by mass of the other elements. The maximum mass of a star with the composition of the Sun is calculated to be about 120 solar masses, if the star is to remain stable. If a star has absolutely no elements other than hydrogen and helium, its maximum mass can be considerably higher than even these estimates – several hundred times the mass of the Sun.

These theoretical calculations are given credibility by observations of the most massive stars known. The most massive is R136a1, which is the brightest star in the dense star cluster R136, the central concentration of stars in the larger star cluster NGC 2070 (entry number 2070 in the New General Catalogue). The cluster is part of the Tarantula Nebula complex in the Large Magellanic Cloud (see page 107), which is made of material with half the proportion of heavy metals that the Sun has. R136a1 has a mass variously estimated at 265–315 times the mass of the Sun. There are some other young stars that are temporarily even more massive than this, some way above 300 solar masses, but they are unstable and losing mass, because it is being blown off by the pressure of radiation.

The first stars were likely therefore to have been very massive and as such they had short lives measured in only millions of years: more massive stars burn hydrogen at a fast rate, and even if they have more of it to burn, they use it more quickly, just as a profligate millionaire might go bankrupt more quickly than a poorer, frugal miser. They became red supergiants and then exploded as supernovae: there are few or no stars of this first generation that survive to the present day. As supergiants, they burnt helium to make carbon, which they ejected as dense clouds of dust that cloaked them and their companion stars in dusty, opaque veils.

Supermassive black holes

Stars are not the only source of the light generated in a galaxy. Virtually all galaxies develop at their heart a supermassive black hole. Black holes themselves are black but as material falls towards a black hole, the energy the material loses is radiated in the form of light and other radiation. The birth and early life of supermassive black holes is a factor in the story of the Dark Ages and the Cosmic Dawn.

Within a month or so of learning about Albert Einstein's general theory of relativity, and even before Einstein formally published it in 1916, the German mathematical physicist Karl Schwarzschild (1873–1916) used it to develop the idea of a black hole. Schwarzschild imagined the following situation: according to general relativity, space curves around a massive body due to the gravitational distortion of spacetime, which causes light to follow curved paths. If a body existed that is sufficiently massive and sufficiently small, then light from the surface of the body might curve so tightly that it might never reach any more than a small distance from the body. The body would then be black, because light would never leave it. The properties of bodies like this were summed up in the name given to them by the American physicist Robert Dicke about 1961 and popularized in 1967 by the theoretical physicist John Wheeler: 'black hole'.

The surface that divides a black hole from the outside world is called the event horizon (pl. VII). No news about anything that happens inside the event horizon can escape through the event horizon: any material (like a newspaper) or any radiation (like radio waves or light) that might carry news about events is dragged back by the strength of the black hole's gravity. The event cannot get across the horizon. Just outside the horizon, gravity strongly bends the tracks of light rays from anything that is visible there and its image can escape but is very distorted.

Although their mathematical properties were quite well developed for more than at least half a century after Schwarzschild's work, black holes remained only a theoretical possibility and had never been observed in nature, so for a long time they were a solution looking for a problem. We now know that nature makes black holes in at least two ways: stellar-mass black holes are produced by supernova explosions (see Chapter 8) and grow by mergers of small black holes into bigger ones; supermassive black holes are produced in the nuclei of nearly all galaxies. There may also be so-called 'intermediate' black holes in clusters of stars, produced by a third, unknown way.

Isolated black holes are dark and difficult to see. However, some stellar-sized black holes exist in binary star systems, which can merge to produce gravitational radiation (see Chapter 8). Additionally, if matter (gas) falls into a black hole, it releases gravitational energy, which heats the gas. This can happen if the black hole has a companion star that leaks gas onto it, or if other stars get drawn near the black hole, break up and then fall into it. These two scenarios make some black holes visible as, on the one hand, X-ray binary stars (see Chapter 8) and, on the other, quasars.

When the early stars exploded, they created black holes of a mass comparable to the stars from which they formed, say ten to one hundred times the mass of the Sun. At about the same time, perhaps even earlier, before the first stars formed, a parallel process began to make another form of black hole, so-called supermassive black holes. These can now be as massive as 1 billion solar masses (the most massive known is Jo313-1806 at 1.6 billion solar masses). They started off smaller and grew by accumulating more and more mass from the parent galaxy.

How small were supermassive black holes when they first started to grow? That is a mystery. Astronomers used to think that they grew from stellar black holes made by supernovae; where there were many stellar-sized black holes, they merged together to make a bigger black hole, which drew in further black holes. However, astronomers now think that this happens too slowly to have worked in time to make the earliest supermassive black holes. There is evidence that supermassive black holes had come into being only a few hundred million years after the start of the Universe: the oldest-known supermassive black hole, Jo313-1806, was in existence just 670 million years after the Big Bang. That is too short a time for 10–100 solar-mass black holes to merge together to grow into a 1-billion-solar-mass black hole – Jo313-1806 would have had to start off at 10,000 solar masses.

If mergers of stellar-mass black holes built up supermassive black holes, they must have got off to a good start via some other process. One likely possibility is that gas and thousands or hundreds of thousands of stars in the central regions of a galaxy formed a cloud that was so dense that it collapsed all together into a medium-sized black hole. This black hole was able to grow to be supermassive by drawing in further stars and clouds, and other black holes.

Some astronomers think that such events in the centres of the youngest galaxies were the triggers that provoked the formation of stars generally in galaxies. Of course, we owe our existence to the Sun. If the Sun was a side effect of the formation of a black hole in our Galaxy, we, too, owe our origin to the action of a supermassive black hole.

Whatever process it was by which supermassive black holes form, it caused an immense pile-up in the centre of the parent galaxies. Packed into a small volume, a black hole of 1 billion solar masses has a gravitational effect that makes it impossible for light to leave, hence, of course, the term 'black hole'.

Quasars

Although supermassive black holes themselves are black, anything that falls down towards the black hole need not be black. If enough material falls, the energy released can be prodigious. Material in the neighbourhood of the black hole is energized and blazes bright in what is called a quasar. Quasars were quickly formed in most – perhaps nearly all – new-born galaxies during the Dark Ages, possibly only millions of years or at most only a few hundred million years after the Big Bang.

The first supermassive black holes drew in the surrounding gas and dust and other smaller stars, converting their gravitational potential energy into radiated energy of various kinds – light, X-rays, radio and so on. Trapped inside the dusty cloak of gas and dust that permeated the parent galaxy, the energy could not escape directly, so at first the Dark Ages remained dark, but the energy warmed the dust in the quasar's parent galaxy, which then re-radiated it as infrared and other radiation. Emerging from the obscurity of the Dark Ages, infrared-radiating galaxies have been detected by infrared- and microwave-sensitive telescopes. (Infrared radiation has a wavelength between about 1 micron to about 1 millimetre, microwave radiation between 1 millimetre to 1 metre, but there are no distinct dividing lines. Because infrared and microwave radiations are absorbed by water vapour, the performance of telescopes sensitive to these radiations is very limited if they are located in a damp environment. They work best if they are located in the dry air on the tops of high mountains and in Antarctica, or are orbiting in space.)

The first stars and black holes having appeared, they poured energy into their parent galaxy. This not only heated the dust, but also pushed it by radiation pressure. The galaxy's interstellar clouds of dust and gas dissipated, dispersing into the outer regions of the parent galaxy and surrounding space. This parted the curtains of dust and revealed the galaxy's stars to the outside Universe and this is the way in which the first galaxies began to end the Dark Ages.

The outflow of energy from stars and black holes in the same galaxies went on to push away material that might be about to fall into the black hole – because the black hole ate so much, it stopped itself feeding. This process is called feedback. The quasar having switched off, however, the gas and dust may then have started to fall back into the galaxy, perhaps accelerated towards the centre of the galaxy and into the black hole by the near passage of another galaxy. This alternation of feast or famine developed into a cycle in a typical galaxy in which its quasar and the formation of new stars switched on and off, alternating every few billion years – but the Dark Ages had ended.

Galaxies that host supermassive black holes reveal themselves in a number of guises. The first type was identified in 1943 by American astronomer Carl Seyfert (1911–1960), at the time a research fellow at the Mount Wilson Observatory in California, although it took decades for their extraordinary nature to emerge. Seyfert noticed their unusually bright nuclei and that they have strong emission lines in their spectra coming from gas that was, in some cases, moving very quickly. The gas was in orbit around something very massive. At the time, the galaxies were enough of an unexplained curiosity to be given a distinctive name: Seyfert galaxies.

In the next development, radio astronomers discovered that some galaxies emit radio waves – the first recorded radio galaxy is on Grote Reber's 1939 radio map of the sky (see page 109). On the map it was confused with the structure of the Milky Way in the constellation of Cygnus and was not at first recognized as a separate feature. In 1946, British physicist J. S. Hey and his colleagues used war-surplus radar equipment to study this source. They named it Cygnus A, the strongest celestial radio source in that constellation. The source was very small and some astronomers thought that it was a new kind of 'radio star'. Others, including cosmologists Austrian-born Thomas Gold and British Fred Hoyle (see page 269), argued that it was not a star, it was outside the Milky Way and some sort of galaxy in its own right. Some years later, they were proved right.

In 1951, University of Cambridge radio-astronomer Sir Francis Graham-Smith (b. 1923) measured the radio position of Cygnus A accurately enough to make it worthwhile to try to find out what was at the same place in the visible sky. The quickest way he could communicate to his colleagues in California, where there were the largest telescopes, was by airmail. German-American astronomer Walter Baade (see page 122) at Caltech took pictures at that position in April 1952 with the 200-inch Hale telescope on Mount Palomar: 'There were galaxies all over the plate, more than two hundred of them, and the brightest was at the centre. It showed signs of tidal distortion, gravitational pull between the two nuclei – I had never seen anything like it before. It was so much on my mind that while I was driving home for supper, I had to stop the car and think.' Baade concluded that Cygnus A was two galaxies in collision.

Discussing his discovery with a sceptical colleague, Ralph Minkowski, Baade bet him that the spectrum of Cygnus A would have spectral emission lines from gas made highly energetic through the collision. The stake for the bet was a bottle of whisky. Minkowski soon took the spectrum with the Palomar telescope, found that Cygnus A did indeed have emission lines and conceded the bet. He need not have proffered the whisky, since the emission lines come not from a collision between galaxies but from gas interacting with a massive black hole, but the whisky was long gone by the time that was clear and he did not get the bottle back.

In the decade following the Second World War, radio astronomy technology improved rapidly and University of Cambridge radio astronomer Martin Ryle invented the powerful kind of radio interferometer for which he received the Nobel Prize in 1974 (see page 22). From 1953 onwards his telescopes were used to make radio surveys of the sky and discovered vast numbers (several thousands) of sources. The sources were distributed evenly all over the sky, so were not likely to be something in our Galaxy; they proved to be galaxies that emit radio waves. One comprehensive and accurate catalogue was published in 1959: sources listed in it received the designation 3C (the third such catalogue made in Cambridge).

The seventh-strongest source in the 3C catalogue had the number 3C273. 'Strong' could indicate that the source was relatively nearby, and if it was a galaxy, as understood up to then, it would be readily visible in sky pictures, but there was no obvious galaxy in the same direction. To be more definite it was necessary to determine the position of the radio source even better than Ryle's telescopes could determine. This became possible because of a piece of luck: from time to time the Moon passes in front of 3C273. British radio astronomer Cyril Hazard used the newly built Parkes radio telescope in Australia to watch a series of such occultations in 1962. He was able to pin down its position by plotting the leading and trailing edges of the Moon at the moment that 3C273 disappeared and reappeared – the source was where the two edges intersected. A number of people noticed that the radio source was near to what looked like an unremarkable star; American astronomer Tom Matthews at Caltech was the first to nail down the position and prove the coincidence. The similarity between the radio source and an ordinary star gave rise to the technical names Quasi-Stellar Object or Quasi-Stellar Radio Source for this kind of astronomical object. The name was a mouthful and came to be abbreviated as 'quasar'.

At first no one could credit that the star was really responsible for the radio emission. But Caltech Dutch-American astronomer Maarten Schmidt (b. 1929) was minded to take a spectrum of the 'star', in order to eliminate it. What he found showed how important it is in science to be methodical and comprehensive and not jump to a conclusion. The star was very bright for the 200-inch telescope and his first attempt was overexposed. He persevered and discovered from a second attempt that 3C273 was *not* an ordinary star. It had spectral lines in emission, indicating hot gas was present. There was some energetic process going on and that could be why the 'star' was a radio source. Trying to be more specific, Schmidt attempted to identify the emission lines in order to figure out what might be happening. The emission lines did not match with anything he had seen before, though he tried several different sorts of explanations. Nor could they be identified by other world experts to whom Schmidt showed the spectrum.

Collaborating with Hazard in writing up all the work on 3C273, Schmidt tried to systematize the wavelengths of the lines in a diagram and suddenly noticed that four of them formed a progression that reminded him of the spectrum of hydrogen – but with the wavelengths redshifted by a huge factor. When he applied the same factor to the other spectral lines, their identification made immediate sense. It all worked out if the spectral lines were redshifted by an unprecedented large amount.

The largest redshifts in astronomy are the result of the expansion of the Universe. If that was the origin of the redshift, 3C273 was receding at record-breaking speed. Schmidt had discovered that 3C273 was not a star, but a galaxy at a huge distance. To look like a nearby star but in fact be at such a distance, 3C273 was radiating energy at a power never before seen. Schmidt had solved one problem by identifying 3C273 but generated another: what was the source of such a power? And that was not all, there was a further problem to come.

On photographs, 3C273 was very small – star-like – but it proved also to be very small in reality. This became apparent in 1963 when American astronomers Harlan Smith (1924–1991) and Dorrit Hoffleit (1907–2007) looked back through the archive of sky photographs at the Harvard College Observatory and discovered that the 'star' that was coincident with 3C273 varied by large amounts on a timescale of years. This meant that it could only be at most light years in dimension, contrasted with the size of a normal galaxy, which is many tens of thousands of light years in size.

Incredibly bright, incredibly distant, incredibly small – it took something incredible to provide an explanation, and it is that quasars are supermassive black holes. Supermassive is not an exaggeration: the most massive black holes in the centres of galaxies have masses of billions of times the mass of the Sun.

Quasars are bright because their black hole is a kind of an engine that turns stars and gas that fall in into light radiation. The light radiation heats up gas and dust that surrounds the black hole. The nearest material to the black hole, in orbit around it in a disc whose size is comparable to the size of the solar system, is gradually spiralling down into the black hole. It gets heated so much that it emits X-rays: many of the 'stars' seen by X-ray telescopes in space are in fact quasars. The power output of quasars is so huge that they can be seen at vast distances. Modern surveys of quasars look out into a huge volume of space: they have identified hundreds of thousands. In 2020, a catalogue compiled with the Gaia space satellite data (see pages 100–102) listed more than 1.6 million. A map of quasars is effectively a map of the Universe out to its most distant parts. Powerful and ubiquitous, quasars initiated Cosmic Dawn and continue to control the Universe.

Colliding galaxies

Brief outbursts from black holes may be individual chance events that are caused by a particular star or gas cloud that ventured too close, the origins of that event usually lost in the mists of the past. There are also larger, more frequent and longer-lasting outbursts when the black hole is repeatedly feasting. They can be triggered by a close encounter between a galaxy and another one that passes by. It may even be that two galaxies collide, which not only wakes up the supermassive black hole in each but causes a starburst in which each galaxy lights up with newly formed, bright stars. Supermassive black holes awakened and bright stars were born: Cosmic Dawn got well under way (see page 82).

As described in Chapter 3, the cosmic web is made of empty voids, but these are surrounded by filaments and clusters, where the galaxies are crowded together. If everything was spread out evenly, galaxies would seldom collide, but because there are extra-dense regions in the cosmic web, galaxies sometimes do collide. The stars in the galaxies are rather small compared to the distances that separate them, so the stars themselves collide only occasionally, the galaxies passing through each other without touching, like marching bandsmen on parade. But, unlike bandsmen, the stars do attract each other by a mutual force of gravity. As the collision proceeds, the galaxies become disturbed in shape. Some stars even get flung out of their parent galaxy, coursing into lonely intergalactic space.

It is interesting to imagine what you would see if you lived on a planet orbiting a star in one of the colliding galaxies. The collision would take place over hundreds of millions of years so it doesn't seem possible that any single individual would comprehend the whole event. It would start with the approaching galaxy looming in a segment of the night sky separated from the normal milky way visible to the planet's inhabitants. Our Milky Way stretches on a great circle around the galactic equator, but the milky way in a colliding galaxy would twist and loop, drawn into a distorted shape across the night sky by the pull of the oncoming intruder. Mid-collision, the entire sky would be bright with stars, obscuring the more distant galaxies. If one's own star suffered the indignity of being ejected from its home galaxy, you would look back to see both galaxies gather together in that part of the sky and gradually recede. The night sky would darken and become free of stars. You would be alone in the dark of intergalactic space for the rest of eternity.

Depending on the circumstances of a collision of two galaxies, they might 'stick' together, and they might merge. The way by which this happens is somewhat similar to the way that brakes work on a car. Just as brakes turn the kinetic energy of the car into heat and stop its motion, so the collision between two galaxies might 'heat' their stars – not by making them individually hotter, but by making them collectively move faster. With the extra speed, the stars can penetrate further in their orbits from their parent galaxy – each galaxy expands. It is not a tidy expansion because the collision would cause the stars to slosh about. Surges of stars make tidal waves in the outer parts of their galaxy. The tidal surges look like shells.

Galaxy collisions are the way that galaxies grow over time. The Hubble Space Telescope set out to investigate how that growth took place. It made deep exposures of two or three areas in the sky, where there were neither stars nor any galaxies known up to that time, to investigate the faintest galaxies that it could see. In astronomy, in general, fainter means further, which in turn means older, so the galaxies that it discovered were very far away, and back a long time ago, close to the time of the Big Bang. These exposures are collectively known as the Hubble Deep Fields (pl. II). They found that, indeed, early in the life of the Universe, galaxies were smaller and more numerous than now. The galaxies had been pictured emerging from the Dark Ages up to half a billion years after the Big Bang. Because they were more numerous than now, they collided more readily and often showed as more disturbed in shape, with tails and arcs of stars at the point of being flung into space. Over the lifetime of the Universe, irregular galaxies like these have merged and settled down to form the larger, more regular galaxies that we see around us now.

Collisions between galaxies not only cause galaxies to merge and get larger, they may be transformational, changing the galaxies' appearance and subsequent history. If a small galaxy collides with a large one, it is called a minor collision. Such a collision 5 billion years ago had, for us, the momentous consequence to conceive our Sun: see pages 105 and 124). However, the small galaxy did not cause much of a disturbance and it was absorbed by the large one without too much drama. On the other hand, if two big galaxies undergo a major collision, it can be transformational, changing the entire nature of the galaxies – for example, spiral galaxies can be transformed into elliptical ones. This is something that will happen to our Galaxy in the future – it will become part of an elliptical galaxy.

There are broadly two kinds of galaxies. Some are flat, rotating systems with spiral arms, like ours. They are the result of the collapse of a slowly rotating intergalactic cloud of hydrogen gas. The cloud rotates faster as it gets smaller and centrifugal force spins the cloud into a flat disc, in which spiral arms appear. This process creates a spiral galaxy.

The beauty of spiral galaxies does not survive in the close company of another. Consider two spirals that rotate about axes that are inclined at a large angle to each other and that are set on a collision course. The first thing that happens as the galaxies approach each other is that their spiral arms are pulled out and the galaxies make grotesque shapes. Then the gas in the two galaxies crashes together and makes especially dense regions. These regions make stars. Because the crash is quick, the combined galaxy shows brightly, with a burst of star formation, using up all the gas. Over time, the two galaxies merge and the stars of which they are made get muddled up. They have no particular way in which they can go, so the merged galaxy does not know which way it should rotate. It loses any spiral galaxy characteristics and develops into an elliptical galaxy. 'Elliptical' refers to its appearance in a two-dimensional picture. It is really a three-dimensional shape, something like a ball. Perhaps it is spherical, but more likely it is ellipsoidal: the shape of a rugby football (a prolate spheroid), or flattened at the poles like the planet Earth (an oblate spheroid). Or perhaps it is even more strangely shaped and has no rotational symmetry at all: it is a tri-axial ellipsoid.

There is more of a chance that galaxies collide if they are in a crowded region. Spiral galaxies that formed in such regions have likely collided already and so, nowadays, elliptical galaxies are more common in clusters of galaxies. The outer regions of clusters and the filaments of the cosmic web are not as crowded, so spirals are more common in these regions of space.

Collisions might be the main way that galaxies grow and evolve but that process, as we understand it, does not explain all galaxies. There are some abnormally massive elliptical galaxies in the central parts of clusters that are difficult to explain as the result of mergers. They seem to be related to galaxies hidden in the Dark Ages that have been discovered by NASA's Spitzer Space Telescope and by the Atacama Large Millimeter/submillimeter Array (ALMA) telescope in Chile (see page 184). Spitzer identified a number of faint galaxies that are not seen with the Hubble Space Telescope's most in-depth view of the Universe, 10 billion light years away, and ALMA was able to study half of them in further detail. ALMA confirmed that they are massive, star-forming galaxies that are producing stars one hundred times more efficiently than the Milky Way. These galaxies are representative of the majority of massive galaxies in the Universe 10 billion years ago, most of which have so far been missed. They are unexpectedly abundant, their numbers well exceeding predictions from theoretical simulations. They assembled during the first billion years of the Universe. It has so far proved impossible to explain how such big galaxies were formed so fast.

Starbursts and quasar outbursts

Collisions between galaxies cause the galaxies to get brighter. This moved the life of the Universe on from the Dark Ages towards the Cosmic Dawn.

During the mergers of galaxies, stars do not collide because they are small and well separated, but clouds of interstellar gas are bigger and do. Gas at the collision interface is squashed. In especially dense regions, gas in the cloud collapses to form stars. If in an image we catch two galaxies colliding, the new stars show as clusters of stars that are bright, hot and blue, with the surrounding gas excited by ultraviolet starlight glowing brightly as a jumble of nebulae. Such an event is known as a starburst (pl. VI). Looking across the whole population of stars in a galaxy, astronomers can identify epochs when starbursts happened, each one triggered by a succession of collisions. The life of a galaxy and its stars is marked by these active episodes.

Shortly after each starburst – perhaps tens of thousands to millions of years after – the newly born stars begin to explode as supernovae, so there is a sudden wave of supernova explosions. There might be as many as one hundred times more supernovae than usual – one per year in a galaxy rather than one per century.

The stars that explode as supernovae are well developed in their evolution and have progressed beyond the burning of hydrogen to helium. They have been burning helium to carbon, carbon to oxygen, oxygen to neon, and so on to magnesium, silicon, sulphur, argon, calcium and iron. The explosion spreads these elements into space where some condense to make dust particles – graphite particles of carbon, sand-like material made of silicate compounds with silicon and oxygen, iron particles and so on. The dust made in the early colliding galaxies was formed in abundance because the first stars were large, evolved rapidly and made lots of the relevant elements before exploding as supernovae. The dust cloaked the light and heat from the remaining stars. It absorbed the light energy and so it became hotter. It emitted infrared radiation, which can be detected by infrared-sensitive telescopes. Microwave radiation with wavelengths in the millimetre range also carries heat from warm things like dust grains: millimetre-wavelength radiation like this can be detected by the ALMA telescope in Chile. ALMA has been able to detect galaxies densely covered by dust so opaque that they cannot be seen at all by the Hubble Space Telescope.

Another effect of the collisions is that the motion of stars and gas in each galaxy is disturbed. Stars and gas no longer orbit the centre of their galaxies in near-circles and the galaxy loses its circular symmetry. If it is a spiral galaxy, it may develop a central 'bar' of stars, with its spiral arms starting at each end of the bar (see page 103). Individual stars and streams of gas become redirected into orbits not only around the galaxy in circles but also passing in and out of the galaxy. This raises the possibility that they will pass near to the galaxy's central supermassive black hole. The pull of the black hole on the nearer side of a star is greater than on the far side and is a tidal force (see page 113). The star may disrupt and its material will join back into the interstellar gas stream flowing onto the black hole.

Black holes do not mind what they eat to make a quasar. Like lions and crocodiles in the Serengeti Park in Tanzania waiting for the tide of wildebeest, impala and zebra on their annual migration, they gorge on anything, usually gas and dust, but 'spaghettified' stars (see page 113) are tasty, too. If such a star drops onto a black hole at the centre of a galaxy, more mass than usual flows towards the quasar in a process called accretion. The flow may often be so great that the black hole cannot swallow it all at once and the material from the disrupted star circles round the black hole in an accretion disc, gradually leaking through onto the black hole itself. The increased mass flow onto the black hole causes the quasar to brighten. The brightening is episodic as extra-big lumps drop into the accretion disc and then into the black hole, so there are short outbursts as well as an overall longer increase of power output. The effects become visible as a bright flaring quasar.

If not all, then nearly all galaxies have a central supermassive black hole. If two galaxies merge, the merged galaxy has two black holes. There are some galaxies, such as NGC 6240, that have a double nucleus, a bright spot of radiated energy coming from two black holes, each drawing in surrounding material. The supermassive black holes are separately eating gas and stars.

In further interactions with the stars and the other black hole, each black hole approaches the other, dining together in close company, then they eat each other. The two black holes gravitate together and orbit each other making a binary black hole pair, stirring up and consuming gas and stars that wander nearby. As they orbit, they radiate gravitational waves. The black hole binary system loses energy and its orbit speeds up, causing gravitational radiation at an increasing power. This lasts for tens or hundreds of millions of years, with the two black holes drawing ever closer.

There is a quasar, OJ 287, which seems to have reached this state. For the last 130 years at least, it has been giving a double burst of light every 11–12 years. One interpretation of what is going on in OJ 287 is that it is a binary black hole pair, so close together that they appear to us as one bright nucleus in a much fainter, merged galaxy. The orbital period of the smaller one around the larger one is twelve years. Twice per orbit the smaller smashes through the accretion disc around the larger, the collision and the brief meal making a double flash.

Eventually the two black holes get so close that they merge in a frenzied, but silent, crescendo of gravitational wave energy. After a very long build-up, the final merger is quick, with prodigious amounts of energy, radiated in a brief burst. The burst could be detectable from right across the Universe, given the right gravitational wave detector. Like radio antennas, gravitational wave detectors are of different sizes according to the frequency range at which they operate: small antennas are sensitive to short waves, namely those of high frequency. Terrestrial gravitational wave detectors like LIGO (see page 172) operate at the wrong frequency range to pick up merging supermassive black holes because the detectors are too small (less than the size of the Earth). Detectors in space can be much bigger (as described on pages 265–66, the space-borne gravitational wave detector eLISA is planned to be six times as big as the Earth–Moon distance) and will be able to see these events.

It seems likely that events from merging supermassive black holes will also produce other forms of radiation, like X-rays and radio waves. The gravitational waves themselves scarcely interact with matter, and little of their energy is directly converted via matter into light and radio waves. However, matter accreting onto the black holes is violently disturbed and radiates in a way that is correlated with the gravitational waves. With attention drawn to a merger event by detection of the gravitational waves, together with some information about the direction from which it comes, derived from the orientation of the detector, astronomers may be able to track the event down to the galaxy in which it appears. Of course, if the galaxy is already recognized to contain two black holes it will be easier to do this. For OJ 287, one theory predicts that the ultimate merger will happen in about ten thousand years, so there is no urgency to get ready. eLISA could well detect between ten and one hundred mergers of supermassive black holes per year, out to a distance of 12 billion light years, with in addition 10 per cent of them detected as X-ray and radio sources. These events provide opportunities for wonderful multi-messenger astronomical investigations that the eLISA scientific teams are planning (see Chapter 13).

Extra energy is liberated by the collision, by the newly born bright blue stars in the starburst, by the supernovae that they make, by the quasar outbursts and by black hole mergers. It is considerably more than normal starlight. It has a profound long-term effect on the development of the merged galaxy that is the outcome from the collision.

Cosmic Dawn

In 1965, soon after Maarten Schmidt had identified the first-known quasar, he gave a talk at Caltech about them. Whatever they were, Schmidt described why the quasars had an enormous redshift and were therefore at prodigious distances in the Universe. This made it possible for astronomers to use them as probes to investigate the intergalactic space across which their light had travelled. There were two PhD students in the audience who noticed something about Schmidt's data that had been overlooked because it was something that was not present in the spectra, rather than something that was. The two students were Jim Gunn (see also page 53) and Bruce Peterson.

Gunn and Peterson saw that short-wavelength ultraviolet light was abundant in the spectra, without any sign that it was being absorbed. Ultraviolet light with a wavelength less than 1,216 angstroms is very readily absorbed by hydrogen atoms and the Universe is full of hydrogen. (An angstrom is a unit of length, conventionally used to measure the wavelength of light, ultraviolet light and X-rays. Named after a Swedish physicist, it is exactly the same as one-ten-billionth of a metre.) Why was the hydrogen that lies between us and the quasar not absorbing the ultraviolet light? It would not be a subtle effect in the spectra of quasars: all the short-wavelength ultraviolet light ought to be absorbed. This analysis came to be known as the Gunn-Peterson effect. It does not often happen that two students yet to be awarded a PhD discover an important scientific effect and have it named after them.

The importance of the effect is that it shows us the state at the present time of the most abundant material in intergalactic space – in fact, in the Universe. As Gunn and Peterson concluded, the likely resolution of the problem is that the hydrogen in intergalactic space is not mainly in the form of hydrogen atoms. The hydrogen atoms have been ionized into their components: electrons and protons. Estimates are that, typically, currently there are fewer than one neutral hydrogen atom in intergalactic gas clouds for every ten thousand free-floating protons, each matched with one buzzing electron.

Even at this low density, the neutral hydrogen atoms do show some effects on quasar spectra, which became apparent as the quality of the available spectra improved. Individual spectral lines, designated as Lyman-alpha, show up in the light of the quasar, imprinted as the light from the quasar travels, through cloud after cloud, to us. The spectral lines appear at the range of wavelengths that correspond to the redshift of the clouds, in the line of sight one after another. The rays of light from the quasar pass through the clouds, threading through them, just as a skewer pushes through meat cubes in a kebab. In distant quasars the spectral lines are so numerous that they look like rows of trees and are referred to as the Lyman-alpha forest. They constitute a demonstration that intergalactic space is full of hydrogen gas clouds. These clouds are the primary constituent of the Universe. Stars, galaxies and quasars get more attention only because they shine brightly and call attention to themselves with louder voices. The Lyman-alpha forest is related conceptually to the Gunn-Peterson effect: the effect is what happens when the Lyman-alpha forest is very dense.

The hydrogen in the Universe at the time of the Big Bang was ionized and as the Universe cooled, it combined into atoms, a process that was completed after several hundred thousand years. The atoms persisted into the Dark Ages. Nowadays, according to the Gunn-Peterson effect, the atoms are broken up again into ions. When did this occur? The question is framed in astronomy as 'When was the epoch of reionization?' and does not have a precise answer. Like the dawn of a winter's morning, Cosmic Dawn developed slowly, at first revealing formless shapes (dust-shrouded galaxies) and then recognizable features of the astronomical landscape (bright galaxies and quasars).

The question can be addressed by looking for the Gunn-Peterson effect in quasars at progressively greater and greater distances, looking back in time into the Dark Ages. Only a few cases have been discovered. They suggest that reionization started when the Universe was about 300 million years old and was complete by about 800 million years.

A similar story emerges from the statistics of the furthest galaxies known. The most distant galaxy as I write is GN-z11 (although there are other contestants for this title, where the evidence is generally regarded as less definite). GN-z11 was found by the Hubble Space Telescope using its infrared capabilities. The galaxy is much smaller than the Milky Way but is very bright for its size because it is forming stars at a great rate and the stars are bright (because they are made of pure hydrogen and helium).

Because it is the most distant galaxy, GN-z11 is the galaxy recorded soonest after the Big Bang, seen as it was when the Universe was 400 million years old. Over the next few hundred million years the clouds of dust that hid younger galaxies gradually dissipated further, the stars emerging into view from within, and Cosmic Dawn was fully achieved some hundreds of millions of years later, ending the Dark Ages. The clearing of the dust also revealed the earliest supermassive black holes shining as quasars. The most distant quasar known at the time of writing was discovered in January 2021 and is called J0313-1806. We see it as it was 670 million years after the Big Bang, just emerging from the Dark Ages into the Cosmic Dawn.

It may well be possible to observe directly the appearance and then disappearance of neutral hydrogen in intergalactic space over the epoch of reionization. Neutral hydrogen has a characteristic spectral line at a wavelength of 21 centimetres, equivalent to a frequency of 1,420 megacycles per second (MHz). Radio telescopes looking back into the distant cosmological past have to tune to a much longer wavelength (lower frequencies) to detect this spectral line. The Experiment to Detect the Global Epoch of Reionization Signature (EDGES), led by American cosmologist Judd Bowman of Arizona State University, is a small ground-based radio telescope located in a radio quiet zone in the Murchison Radio-astronomy Observatory in Western Australia. It looks at the sky as it passes overhead, and it has to tune to radio frequencies of 50–100 MHz (wavelengths of 6.0–3.0 metres) to see 21-centimetre radio waves from neutral hydrogen at the right time in the past. At the longest wavelengths (biggest redshifts) it sees no signal from cosmological neutral hydrogen. As it tunes in towards the reionization, it sees the signal from neutral hydrogen turn on as the Universe cooled, and then turn off as the proportion of neutral hydrogen diminishes again in the reionization.

According to first results from EDGES, published late in 2019, the first light of Cosmic Dawn happened 250 million years after the Big Bang. This estimate for the time of Cosmic Dawn is a bit discordant with other methods and more work will be needed to come to a consensus that reconciles all the different methodologies and whatever it is, precisely, that each one measures.

Cosmic Noon: when the Universe was brightest

Cosmic Dawn, the epoch of reionization, started when the Universe was about 300 million years old and was completed mostly by 600 million years, certainly by 800 million years old or so. Since then, the Universe has been, in overall appearance, the Universe that we see until today: a universe of galaxies, shining into clouds of ionized hydrogen. What happened to the Universe after Cosmic Dawn was the equivalent of a human being growing up – a childhood and adolescence, followed by a longer period in which the Universe was the equivalent of a mature and ageing adult, which is the time in which we now live.

The entirety of the history of star formation in the Universe was first elucidated successfully by the Italian astrophysicist Piero Madau in 1996. After Cosmic Dawn, the rate at which stars are born became more and more frequent. The birth of new stars in the Universe reached its maximum rate at 3.5 billion years after the Big Bang: sometimes this epoch, when the Universe of galaxies was at its most mature, is referred to as 'Cosmic Noon'. From Cosmic Dawn to Cosmic Noon, galaxies on average got larger and brighter, with numerous brilliant stars and glowing nebulae. The Universe was at its brightest, aglow with light.

Since Cosmic Noon, starbirths have become less and less frequent, halving every 2.6 billion years as the cosmic afternoon has progressed. As the older stars fade away, there are fewer stars to replace them. As a result, on average, galaxies are becoming fainter. The majority of brilliant, young blue stars that shone into intergalactic space and caused the nebulae of their host galaxy to glow have died in increasing numbers. They are being transformed into fainter red stars, with no new generation of stars to replace them. The reason for this change is that star formation has been diminished by feedback mechanisms, the energy radiated by super-massive black holes and supernovae dispersing interstellar hydrogen gas (see pages 118–19), which, in any case, is less abundant, having been depleted by being used up by earlier stars. At the present time, galaxies are more likely than before to be quenched or starved.

As a result, galaxies are becoming redder and deader. At present, the stellar birth rate on average in the Universe has fallen to one-tenth of the value it had at maximum, and will fall further. After 100 million million years have passed, no more stars will form. Cosmic Dawn was followed by Cosmic Noon and the cosmic after-noon will be followed by cosmic night. The present and likely future states of the Universe are clear, because it is relatively easy to study the Universe nearby: it is gradually fading away, ageing gracefully. It is on its inevitable trajectory towards darkness and cold.

It is more difficult to look back to see the detail of the time when stars first formed, and beyond that to see into the Dark Ages. It will take new telescopes, like the James Webb Space Telescope, which is optimized to record infrared radiation, to clarify the situation and address the issue further. Its work will be supported by specialist radio telescopes currently under construction like the Square Kilometre Array (SKA) and gigantic optical telescopes, such as the European Southern Observatory's Extremely Large Telescope (ELT), and a couple of slightly smaller telescopes operated by American agencies. They will look into the Dark Ages, enabling astronomers to view the Universe during its childhood.

5
Our Galaxy: Birth and Cannibalism

Why does our Galaxy look the way it does? Determining this is akin to discerning the history of an archaeological site. The basic data are the locations, motions and ages of the stars from which the Galaxy is made, some of which show subtleties that indicate their alien origin. Like archaeological finds, they indicate how the site developed, through a succession of collisions and mergers, much as an army might have invaded a historical location, leading to the assimilation and naturalization of the civilian population that migrated after them.

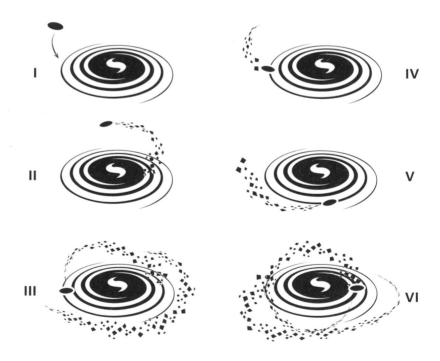

Galaxies merged, building each other up in a hierarchy to progressively bigger sizes. The galaxies that orbited and then merged with our own Galaxy left trails of their alien stars.

The birth of our Galaxy

The Milky Way Galaxy is our galaxy of stars, including our Sun. It is common to call it simply 'the Galaxy', with a capital letter. Although it is clear that the Galaxy is very old, it has been surprisingly difficult to say accurately how old it is. The way that this has been attempted historically is to track down the oldest stars in it: the Galaxy must be at least as old as them. One problem with this method is that, even if you find an old star, you do not know whether there are stars somewhere else in the Galaxy that are older, or whether the star is an interloper having dropped into the Galaxy from outside. However, most of the mergers of other galaxies with our own Galaxy took place early on in its history, so that finding such stars is a promising way to address the problem. A separate but even more difficult issue that immediately follows the discovery of an old star is that the various ways that the ages of stars are estimated are not very precise.

The stars that are easiest to study in detail in an attempt to determine their age are the brightest ones – the ones we see in the night sky. Being bright, they are by and large the nearer ones. They live in the same neighbourhood as our Sun, meaning that they live in the galactic disc. Disc stars are predominantly young. It takes astronomers quite an effort to pick out the few old stars that are coursing through them. In general, the stars of the more distant galactic halo are the remnants of galaxies that merged with our own Galaxy billions of years ago and which have disintegrated. That is the least confusing place to seek out the oldest stars.

The galactic halo is where globular clusters of stars live. By contrast with individual stars, globular clusters are among the most readily recognizable objects in the Milky Way. They are dense clusters of perhaps hundreds of thousands of stars, which are fossils that hark back to the Galaxy's early history. Globular clusters are so called because they are almost spherical. They look like small galaxies, which in a way they are. They contain many stars and they are often at great distances, on the far reaches of the Galaxy, almost in intergalactic space. One is so well populated with stars that even though it is not at all the nearest, it is the brightest. At 15,800 light years, it is so distant, and therefore appears so small, that it is usual for anyone with normal eyesight but no telescope to see it as a star. Accordingly, it was originally given a name appropriate to a star: Omega Centauri.

Omega Centauri is a stunning sight in any telescope, the finest globular cluster in the sky. It is not usually permitted to use an expensive, large telescope for something as frivolous as stargazing but once, waiting for some problem to be fixed so that I could get on with my scientific programme, I looked at Omega Centauri for purposes of astro-tourism with the 4-metre (150-inch) Anglo-Australian Telescope. I took the detector off the telescope's camera and, seated in the observing chair, I looked at the cluster's stars hovering in the focal plane of the camera, over my knees. They were suspended in space, as they would be if one could cruise past them in an interstellar spacecraft. Small tremulations of the atmosphere caused the impression that they were alive, that I was really close to the cluster. I could readily pick out stars of different colours, particularly the bright, red giants. It was a mesmerizing, breathtaking sight.

Since they are so old, globular clusters evidently persist for a long time, but they do not last forever. The dense inner regions are tightly packed and do indeed last for eons. However, stars in the outer regions are less strongly bound to the globular cluster by its gravitational force and they may wander off as the cluster orbits around and through the Galaxy, subjected to random disturbances. Omega Centauri is a glorious sight now but it was even more glorious before it lost its outer fringe of stars, like Samson's hair before Delilah cut his locks off.

Some of Omega Centauri's lost stars have been identified mingling with our own Galaxy's stars – 309 of them flow in a stream called Fimbulthul (named for one of the eleven rivers that, in Norse mythology, flowed through the primordial void). The trail of stars arcs over 18 degrees in the sky in orbits that take them as close as 5,000 light years to the centre of the Galaxy and as far as 21,300 light years from it. Their orbits follow that of Omega Centauri. It seems likely that Omega Centauri is the central, densest part of a dwarf galaxy that has fallen into our own Galaxy and is shedding its stars in the Fimbulthul stream. Omega Centauri is thus not native to our Galaxy, it is an immigrant that settled here.

The age of globular clusters can be estimated by analysing the old stars that they contain. Some of these stars are so-called white dwarfs. Most stars that have completed their lives die by becoming small, faint white dwarfs (see Chapter 8), which are at first hot (white hot – this is the origin of their name). They then cool by emitting radiation – not only light and heat but also neutrinos. The temperature of the white dwarfs gradually drops. I have seen it portrayed in western movies, and I suppose it is true, that a Native American scout can estimate when a campfire was abandoned by feeling the temperature of the dying embers. In the same way, the age of a white dwarf can be estimated from the temperature of its surface. White dwarfs are being made throughout the history of a given globular cluster, and the age of the coolest white dwarf in a cluster is a lower limit to the cluster's age. Exploiting the logic of this story, in 2004 Canadian astronomer Harvey Richer estimated the age of the globular cluster M 4 as 12.7 billion years, give or take 700 million years.

Another way to estimate the age of a globular cluster is to fit theoretical calculations of the evolution of stars to the pattern of the brightness and temperature of the cluster's stars. This gives a range of ages for a number of globular clusters. They have quite a spread, with a likely value for the oldest cluster of 12.6 billion years.

Although it is difficult to find in the Milky Way a very old star that is drifting among all the others without signalling its age by being a member of a globular cluster, the way to do this is to survey the spectrum of lots of stars using a quick technique to pick out the metal-poor ones. 'Metal-poor' means that they have rather small amounts of those elements that have been made progressively throughout the life of the Universe by generation after generation of earlier stars. It has in recent times become possible to do this by making rapid surveys of millions or even hundreds of millions of stars using automated telescopes. This produces lots of data but, with techniques that have fed into more general computer analyses of Big Data used for forecasting in industry and politics, the field can be narrowed to a short list of the oldest-looking stars. More detailed studies confirm which ones really are very old – this is a time-consuming process: you have to kiss a lot of frogs before one turns out to be the prince that you are hoping to find. The oldest found in this way is the star 2MASS J18082002-5104378 B. The complicated designation hints that this star is one of many that are superficially similar but that proved to be an exceptional, although tiny, ultra-metal-poor star born about 13.5 billion years ago.

Another way to determine the age of stars that can in principle be very precise is to look at elements like thorium (Th) and uranium (U) to see how much of each a star contains. These two are long-lived radioactive elements, which gradually decrease with time at a very precisely calibrated rate. The amount that stars contain, compared to the amount expected by looking at every other element that they have, shows how much time has passed. Although this does not carry the uncertainties of theoretical calculations, the method does suffer from the problem that measuring the quantity of suitable elements in the spectrum of a star is difficult because there is usually not much of them there. However, Th-232 and U-238 are the two useful radioactive elements for this purpose because they take a time to decay that is comparable to the age of the Galaxy and leave relatively large traces of their presence in the stellar spectrum. Th-232 has a half-life of 14.05 billion years and U-238, 4.468 billion years. The process of working out when astronomical things happened using nuclear techniques is called nucleocosmochronology. Applying this polysyllabic method to the star CS 31082-001 yielded a value for its age of about 12.5 plus or minus 3 billion years, and the age of BD +17° 3248 came in at 13.8 plus or minus 4 billion years.

All these figures for the age of the Galaxy are uncertain and somewhat difficult to reconcile and astronomers still have more to do to be able to be clear about this topic. But the bottom line is: our Galaxy is indeed very old, having been born, probably, about 13 billion years ago. This was within the first half a billion to a billion years of the Universe, towards the end of the Dark Ages.

The shape of our Galaxy

The Galaxy formed from a gas cloud within which the oldest of its stars were condensing, and took up a distinctive spiral shape. It was not easy to determine the shape of our Galaxy because we are inside it and cannot view its overall structure, but here is how it was done.

As I look out from the upstairs study of my house through a window, I can see some of the city in which I live. Other houses are spread out horizontally all around. I infer that the city is built on flat land – the houses are distributed in a plane that extends out towards the horizon. I cannot tell the size of the city – the houses stretch out into the distance, but beyond a certain distance the nearer houses obscure the houses that lie beyond. I can estimate the distance of the furthest houses that I can see but I do not know how far houses continue. I know that they peter out at the city's edge but I cannot see it. I can see that the nearest houses, including mine, form a row, and I can see that there is another row of houses running parallel to mine. I would find it difficult to map the distribution of the more distant houses, although I could use a laser ranger to find the distance of some of the chimney pots that I can see across the roofs and plot their distribution on a map to find how they line up along other streets in the neighbourhood. If I raise my eyes above the horizontal, I can see some far houses clustered in the distance – there is a village built on a hill in that direction. I can see that in some areas of the village the houses form lines. I can infer that my city has features in common with that distant village – houses are arranged in streets. I could make observations like these to build up a map of my city.

These observations are similar to those that astronomers have made through history to map our Galaxy and show the distribution of its stars, in part from investigating the properties of our

Galaxy itself and in part from looking at galaxies that are like ours. The Milky Way is one of the most obvious features of the night sky when viewed from a site without interference from artificial lights. It consists of large numbers of stars, massed into a filmy band. So, the first clues as to the shape of the Galaxy were from simple observations of the shape of the Milky Way as it arcs across the night sky.

The earliest clear scientific description of the Milky Way was by Ptolemy, a Greek astronomer who lived in Alexandria in Egypt in the second century CE. He was probably a librarian at the famous library of that city and used his access to its collection to compile textbooks, including a book on astronomy now known as *Almagest*, which through Arabic translations became the definitive work on that subject for almost a thousand years. 'The Milky Way is not simply a circle,' he wrote, 'but a zone having almost the colour of milk, whence its name. It is not regular and ordered but different in width, colour, density and position.'

The explanation for the Milky Way's milk-like appearance proved to be that it is made up of many stars that are too faint and close to each other to be viewed individually. This was first conjectured in the fifth century BCE by the Greek philosopher Democritus and proved by Italian astronomer Galileo Galilei (1564–1642) with his telescope in the winter of 1609–10. Galileo wrote in his 1610 treatise *Sidereus nuncius* (The Starry Messenger) that the Milky Way 'is, in fact, nothing but a collection of innumerable stars grouped together in clusters. Upon whatever part of it the telescope was directed, a vast cloud of stars is immediately presented to view.' However, Galileo offered no explanation of the shape of the Milky Way as a band along a great circle of the celestial sphere.

One of the first explanations to get close to a scientific, geometrical model relating to the distribution of the stars was by William Stukeley (1687–1765), an English antiquarian who studied Stonehenge, as well as astronomical phenomena. In his *Memoir* (1757), he suggested that nearer stars – the ones that we can see individually – formed a bounded spherical cluster, surrounded, Saturn-like, by a flat ring of stars. It is this ring that, viewed from a central point where we are located, we see as the Milky Way, thought Stukeley.

The start of modern ideas about the structure of the Milky Way was a model proposed by Thomas Wright (1711–1786), an eccentric, astronomer, mathematician and garden designer from northeast England. In a short book published in 1750, Wright explained the structure of the Universe, starting with the solar system and extending to the stars. A mixture of the profound and the peculiar, the book has more significance than its content might otherwise deserve because the German philosopher Immanuel Kant acknowledged it as the source of his ideas about the origin of the Milky Way, which he relates to other celestial bodies.

Wright's model of the Milky Way proposed that it is a slab of stars seen from the inside. Look from within the slab, in the plane, and many stars and much starlight can be seen. Look across the slab and fewer stars are seen, and thus less starlight. He went on to tweak his model by supposing that the slab of stars was formed by two concentric spherical surfaces of large diameter, so that locally the distribution could appear planar, as far as it could be perceived within a short distance. Wright concluded by imagining how the Milky Way Galaxy is one of a collection of spherical star systems that stretch without end into space, the local region being something 'which you may if you please, call a partial View of Immensity, or without much Impropriety perhaps, a finite View of Infinity'.

It is not too much of an exaggeration to say that Wright's model presages the modern view of the Milky Way Galaxy as a flat collection of stars surrounded by other similar galaxies that extend far into space.

Through a report in a newspaper, Wright's ideas inspired the German philosopher Immanuel Kant (1724–1804), which he acknowledged in his 1755 *Allgemeine Naturgeschichte und Theorie des Himmels* (Universal Natural History and Theory of Heaven), adding:

[Herr Wright] first gave me cause to regard the fixed stars not as a scattered milling mass without any visible order, but rather as a system…extending through the entire heavens, and where they are most densely massed, they form the bright band that is called the Milky Way. I have become convinced that, because this zone, illuminated by countless Suns, has very exactly the direction of a very large circle, our Sun must also be very close to this large plane of reference.

In the second part of his book, by far the longest, Kant presents his nebular hypothesis, explaining how these features of his model originate in the rotation of a large cloud of gas, a nebula. He went on to suggest that the Milky Way formed from a (much larger) rotating cloud. Kant related this idea to the cloudy structures, or 'nebulae', being discovered by eighteenth-century astronomers and eventually catalogued by French astronomer Charles Messier in 1771 and 1781 (see page 58). They might also be similarly large and distant discs of stars.

The question of the relationship between the Milky Way and the nebulae was studied in the late years of the eighteenth century by British astronomer William Herschel (1738–1822). He developed the field of stellar statistics, counting the number of stars that he could see in various directions through his Twenty-Foot Telescope (the size was the focal length of the light-collecting mirror, not the diameter). He called this process 'star gaging', using eighteenth-century spelling. He sampled the population of stars in hundreds of directions, then plotted the results of his stellar census along a great circle in the sky orthogonal to the line of the Milky Way, showing the numbers in a polar diagram. The resulting figure was three to four times longer than it was broad.

In analysing the result of his observations, Herschel assumed that all the stars were more or less the same brightness and formed a uniform distribution in space, and that his telescope was capable of viewing out to the edge of the stellar system and beyond. The numbers of stars thus represented the distance to which the system of stars extended. On these assumptions, the polar diagram was a section through the system. It was flattened, a shape not unlike many nebulae that Herschel had also viewed with his telescope: we now would call these nebulae edge-on spiral or elliptical galaxies. Herschel interpreted the distribution of the stars as a slab, with the Sun on its centre line, and the Milky Way as the effect of our view from within the slab. Herschel noted the bifurcation of the distribution, where the Milky Way is split along its length in the constellation of Cygnus. It is now known that this is the result of obscuration of more distant stars by the dust clouds along the central plane of the Milky Way (pl. 1), but Herschel interpreted it as a lack of stars where the slab was split in two. He related this to the variety of shapes of the nebulae that he had seen.

When in 1789 Herschel completed his even larger telescope, the Forty-Foot Telescope, he discovered immediately that it saw more stars than the Twenty-Foot Telescope, so that it was clear that he had not previously penetrated beyond the edge of the Milky Way stellar system. The distribution of stars extended indefinitely into space, beyond the reach of his telescopes, and his polar diagram was not a representative cross-section through the Galaxy. If anything, the diagram disproved one of his key assumptions, that the distribution of stars was uniform. Nevertheless, Herschel's picture continued to be reproduced, even up to the present day, as if it was a model of the Galaxy, confirming in many ways the view of Wright and Kant. In the closing years of the eighteenth century, Herschel himself continued to hold to a qualitative view that the Milky Way Galaxy was a flattened distribution of stars, analogous to other nebulae. This model has prevailed.

In parallel to his work on the distribution of stars in order to determine the shape of our own star system, Herschel set out to inspect the known nebulae and to find and classify new ones. His method was to sweep his telescope over the sky in parallel, overlapping strips. He noted the features that passed into the field of view that he saw from his perch at the eyepiece, high on the telescope in his garden. He shouted the details to his sister Caroline, who took notes at a table on the grass below or at the first-floor window of their house. She also undertook the task during cloudy nights and the daytime to make systematic, formal catalogues of the discoveries.

Herschel found more than two thousand nebulae, which often looked like flattened discs, and speculated that such nebulae could be Milky Way systems. However, the question of the relationship of the Milky Way to the nebulae remained in doubt for a further hundred years until the astronomers of the early twentieth century concluded that many nebulae were star systems outside the Galaxy, at first called 'island universes' and then, by analogy with our own star system, 'galaxies'. With the invention of photography and its application to astronomy, galaxies were definitely identified as, typically, flattened rotating discs of stars. The stars are orbiting around a central bulge of old, red stars with bright, blue stars arranged along spiral arms, the whole thing embedded in a larger approximately spherical halo. The history of the origin of the Galaxy thus starts with an explanation for this distinctive shape.

The Galaxy originated as a small, chance accumulation in the Big Bang material of hydrogen, helium and dark matter. Over the first few millions of years of its life, the Galaxy grew rapidly more massive, accumulating further dark matter and gas through the accretion of that jumble of material onto the lump. At first the dark matter and the ordinary matter stayed intermingled during the formation of the Galaxy, but the ordinary matter began to cool down relative to the dark matter. This was because ordinary matter radiates energy, whereas dark matter does not. This meant that ordinary matter (the gas) slowed its motion, whereas the dark matter did not. The ordinary matter contracted into the centre of the lump, but the dark matter retained its shape as a larger halo. The dark matter halo of our Galaxy is extensive – it extends out to a distance of 200,000 light years or more – and is typical of all galaxies: the dark matter is more widely distributed than the stars and gas.

Meanwhile, the Galaxy continued to accrete material. At the same time that it grew in mass, it shrank in size and, as a result, it rotated more quickly. The origin of the rotation was in the original lump, which was slowly rotating. As the Galaxy shrank, it rotated faster. The reason is the same one that enables an ice skater to pirouette faster: she makes herself effectively smaller by drawing her arms into her sides and then moving them vertically up above her head. When this happens, her skirt flares outwards due to effects of centrifugal force. Likewise, the gas in the Galaxy formed a kind of skirt, changing from a slowly rotating, more or less spherical shape to a rapidly rotating disc. The dark matter and the gas that first condensed into stars remained in the spherical halo of the Galaxy, but later generations of stars, such as the Sun, formed in the gas of the flat disc. The flat disc of stars is what we see edge-on in the night sky when we view the Milky Way. It is embedded in the halo.

The first mergers with our Galaxy
In the first years of the Big Bang, when the Universe was more tightly packed than now and there were more galaxies because galaxies were smaller, collisions between galaxies were more common. Smaller galaxies merged with larger ones. Our Galaxy grew by galactic cannibalism, eating other galaxies. Some of the earliest acquisitions were galaxies that contained globular clusters, which survived the merger. These alien globular clusters look more or less like the native globular clusters born in our own Galaxy, but they have characteristics, such as their composition, that mark them as interlopers. As many as a quarter of the 150 globular clusters in our Galaxy are intergalactic aliens, now settled in our Galaxy as immigrants.

While compact globular clusters of alien stars survived the merger of their parent galaxy with ours, other infalling galaxies were completely disrupted, their stars now mingling individually with native-born galactic stars. Some mergers have proceeded in fits and starts. The so-called Virgo Overdensity is the heart of a dwarf spheroidal galaxy that fell into the Galaxy in a head-on collision, plunging radially in towards the galactic centre about 2.7 billion years ago. It barrelled right through and out the other side, gradually shedding stars even as it disturbed the Galaxy's stars and began to fall back. The multiple collision has created shells of stars in the halo of our Galaxy. Even now, some smaller galaxies are falling into our Galaxy, leaving behind streams of individual stars and star clusters that orbit in its outskirts. Some streams encircle the Galaxy more than once as the incoming galaxy orbits repeatedly.

Stars get pulled from the infalling galaxy by our Galaxy. These alien stars leave behind a track akin to the condensation trail of a high-flying jet aircraft, marking where the incoming galaxy fell. Like a condensation trail, the star stream may drift slightly from where it was deposited and widen over time as the track diffuses. Over the succeeding millions of years, the stream will have disintegrated, the alien stars intermingling with the rest.

One statistical study suggests that over the course of its lifetime, the Galaxy has cannibalized about five galaxies with more than 100 million stars, and about fifteen with at least 10 million stars (these figures refer to mergers that left identifiable traces). As well as these major meals, the Galaxy would have snacked on numerous smaller galaxies.

The first mergers between our Galaxy and others are unclear from the present available evidence. This is for several reasons. The Galaxy was smaller than it is now, having accreted a number of galaxies since it was born, and grown bigger, so the effect of the mergers was major and repeatedly transformed the nature of our Galaxy. Moreover, much has happened to our Galaxy since the first mergers happened, so the early history is very indistinct.

The earliest merger of a galaxy with ours from which some features have been identified from statistical studies took place 11 billion to 9 billion years ago, when the Galaxy was four times less massive. It must have truly transformed what our Galaxy looked like at the time. Traces of the infalling galaxy were found independently by two groups of astronomers in data from the Gaia space satellite.

Gaia is the name of an ESA satellite launched in 2013 from the European Spaceport in French Guiana in South America. GAIA was originally an appropriate acronym for the spacecraft, but the whole design concept changed during development so the acronym is no longer appropriate; however, the name was kept for continuity.

The spacecraft, which is, as I write, in what will probably be its last years of operation, repeatedly observed a billion stars and other objects (like asteroids) over its lifetime, estimated at nine years, gathering information about their position, their motions and their intrinsic properties. It started observations three weeks after launch at its observing station, a point known as L2, located 1.5 million kilometres (930,000 miles) from Earth in the direction away from the Sun, where the combined pull of the Sun and Earth keep the satellite in orbit, tracking alongside the Earth in its annual circuit. The heart of the satellite consists of very precise and delicate optical instruments, always shielded from the Sun by a heatshield that nullifies temperature variations. The whole arrangement makes Gaia a phenomenal data-gathering machine.

The satellite slowly rotates, four times a day, and has two telescopes that point in directions that are 106 degrees apart. The telescopes feed CCDs, which record what is seen. The 106 CCDs in the ensemble of instruments have a total of 1 billion pixels – an ultra-high definition (UHD) TV camera has just under 9 million pixels, a hundred times fewer. The telescopes scan a circle on the sky, one telescope following the other, and the instruments in the focal plane of the telescopes time the passage of stars in the fields of view. As time progresses and the satellite continues on its orbit, it scans strips of sky that are side by side, repeatedly viewing the entire sky. The amount of data that flows from the telescopes is prodigious. From L2, Gaia transmits with a wireless power of only 300 watts to ESA ground stations in Australia, Spain and Argentina at a rate of up to 10 million bits per second, comparable to domestic high-speed fibre-optics broadband connections to the Internet. A total of some 100 terabytes of science data is being collected during Gaia's lifetime, with the estimated total data archive surpassing 1 petabyte, roughly the same size as the data in all the research libraries in the USA put together.

Over its lifetime, Gaia will observe each of its billion stars about eighty times. The orbit and orientation of the satellite are continuously monitored to a very high precision, and the timings effectively provide the positions of the stars. Over time, changes in the position of the stars provide their distance and their motion across the sky. Further optical instruments and CCDs record the brightness and spectrum of the stars and their velocity towards or away from us. Because Gaia gathers fundamental data for so many stars, it is a wonderful set for astronomers to analyse. In the early years after data was first released, Gaia produced more than fifteen hundred astronomical papers a year. This numerical performance indicator does not tell by any means the whole story about scientific value, but the Gaia team takes pride that it puts its satellite on a par with the hitherto most productive astronomical facility, namely the Hubble Space Telescope.

This gigantic set of data for so many stars makes it possible to identify and examine the statistical properties of different groups of stars in our Galaxy. Stars in the plane of the Galaxy move differently from stars in the halo – they have circular orbits around the Galaxy and orbits that plunge in and out of the Galaxy, respectively.

A group of stars discovered in Gaia data by Russian-born astronomer Vasily Belokurov and his collaborators at the University of Cambridge in 2019 has been given the inelegant name of 'the Sausage' because of the shape of the distribution of the stars in diagrams showing their compositions and their orbital speeds and positions, as measured by Gaia. According to Belokurov and a second group of astronomers, mostly Dutch-based and led by Argentine Amina Helmi, these stars occupy the inner halo of our Galaxy and are the remains of a galaxy about the size of the Small Magellanic Cloud that collided headlong with our Galaxy approximately 11 billion to 9 billion years ago, when it had an age of a few billion years.

Helmi and her colleagues christened the infalling galaxy Gaia-Enceladus, a name intended to be more poetic than 'the Sausage'. In Greek mythology, Enceladus was one of the Titans, and the offspring of Gaia and Uranus (Earth and Sky), said to be buried under Mount Etna in Sicily and responsible for earthquakes. Likewise, the galaxy is an intellectual offspring from the spacecraft Gaia and the astronomical sky; it was a giant galaxy compared to other satellite galaxies of the Milky Way; it has been buried (disrupted by our Milky Way Galaxy and hidden in the Gaia data); and it was responsible for what might be termed seismic activity, or shaking the Milky Way. The name of Enceladus having already been taken by one of Saturn's moons, the galaxy was given a double-barrelled name of Gaia-Enceladus, the authors said, to avoid confusion, even if this adds six syllables and thirteen letters of erudite classical mythology to astronomy. My guess is that, like the first asteroid, Ceres, discovered in 1801 and originally named Ceres Ferdinandea, its name will be shortened in usage.

The development of spiral arms

After the hurly-burly of the early mergers, the Galaxy was able to settle and develop and maintain an orderly structure. The hydrogen gas in the disc of the Galaxy and the stars that formed in it developed a very definite structure: they took up a spiral shape (see page 77). It is very hard to see the spiral from our vantage point in the Galaxy – it is always difficult to see the shape of something from the inside. Nevertheless, it has proved possible to put together a map that reveals that two main spiral arms emanate from the central region of our Galaxy, each twisting through a complete revolution. There are two less dense spiral arms that lie between them.

The spiral arms of our Galaxy do not originate right in its centre. The central region of our Galaxy has the shape of a 'bar' and the Galaxy's spiral arms emanate from the end of the bar (pl. VIII). Barred spiral galaxies like this are common – just one-third of spiral galaxies have arms that originate at their central point, while perhaps two-thirds of the spiral galaxies nearby to us are similar to ours in having a bar. The stars in the bar are older stars and they orbit in and out of the Galaxy.

The reason why the bar forms, and why not all spiral galaxies have one, is something to do with the interaction of the stars in the disc with the galaxy's dark matter halo, but it is a complicated problem and astronomers are not agreed on exactly how it all works.

However, once a bar formed, waves in the gas in the disc of the Galaxy caused the gas to line up in spirals that emanate from the ends of the bar. Stars were pulled into the massive concentrations of gas along the spirals and new stars formed because the gas was so concentrated. New stars are blue and emit energetic ultraviolet light, which splits apart the hydrogen atoms in their neighbourhood into protons and electrons. If the protons and electrons recombined to reassemble the hydrogen atoms, they emitted red light known as H-alpha. The clouds of gas became visible as red nebulae, centred on the ultraviolet-emitting stars and delineating the spiral arms. The pattern of spiral arms, blue stars and red nebulae is very distinctive, and very beautiful, and the reason why colour pictures of spiral galaxies are widely reproduced.

The Hubble Space Telescope makes it possible to look at the development of spiral arms in general, in typical spiral galaxies over most of the lifetime of the Universe. A similar sequence of events must have taken place for our Galaxy. The most distant galaxies, which we see in their infancy a few billion years after the Big Bang, were somewhat shapeless. They had bright, clumpy star-forming regions, similar in their content to spiral galaxies, but without spiral structure. Over the next billion years or so, these galaxies began to develop a more structured appearance, with a central bulge. The effects of rotation began to appear, like the start of the rotation of water as it flows into the drain hole of a bath. Two clear spiral arms started to appear when the galaxies were 3 billion to 4 billion years old, with more complex, multi-armed structures, as we see in our Galaxy, appearing many billions of years later, at the time of the formation of the thinnest parts of the disc, at 9 billion years old.

The stars in the disc of the Galaxy orbit around its centre, but the spiral arms remain stationary – they are self-perpetuating. The reason for this is similar to the reason that congestion persists on a motorway after some sort of slowdown (perhaps for a minor accident). The slowdown starts with the accident itself, building congestion, then, even after the accident is cleared, cars enter the congested area, slow down to pass through it and whizz off once it is behind them. The congested spot remains fixed at a particular point on the motorway all day, even though the cars on the motorway drive through. Likewise, stars move through the spiral arms of the Galaxy but the arms stay fixed. However, gas compressed in the spiral arms changes into stars, so the arms gradually became depleted. The Galaxy is gradually becoming more anaemic, its arms becoming thinner with age.

Accretion of smaller galaxies: mergers and streams

Individual stars, star clusters and small galaxies continue to fall into our Galaxy even now. About a dozen star streams have been identified, left behind from recent infalls. The Field of Streams is a patch of sky where several stellar streams cross. The main one, identified on the celestial sky in 1997 by Belokurov and his colleagues, is the Sagittarius Stream – a thin ribbon of stars that wrap in a spiral around the Galaxy more than once, which shows as a doubling of the stream.

This stream has been strewn from the Sagittarius Dwarf Elliptical Galaxy, which is not a big player – as the word 'dwarf' in its name indicates, it is a small galaxy (our Galaxy contains perhaps 100 billion stars; the Sagittarius Dwarf Elliptical Galaxy contains hundreds of millions – many hundred times fewer). It is 70,000 light years away on the far side of our Galaxy in the direction of Sagittarius, a constellation that is large and full of bright stars and extensive clouds of stars that belong to the bulge in the centre of our Galaxy. The stars of the dwarf galaxy are thus confused by foreground galactic stars and it takes a lot of work to separate them out. This is the reason why it was only discovered in 1994 by University of Cambridge astronomers Rodrigo Ibata, Mike Irwin and Gerry Gilmore.

The Sagittarius Dwarf Elliptical Galaxy is one of the two nearest galaxies to ours. As proposed in 1995 by British astrophysicist Donald Lynden-Bell and chemist Ruth Lynden-Bell, it is orbiting around and through our Galaxy. It fell into our Galaxy 5 billion years ago and has orbited the Galaxy several times. The last two and a half orbits have left their traces of the stars that it discarded on the sky as the Sagittarius Stream.

The Sagittarius Dwarf Elliptical Galaxy has passed up and down through the plane of our Galaxy three times, the first time during its infall 5 billion years ago, again 2 billion years ago and a third time 1 billion years ago, according to astronomer Tomás Ruiz-Lara of the Instituto de Astrofísica de Canarias. When his team looked into the Gaia data for the Milky Way, they found three periods, each lasting hundreds of millions of years, when the birth rate of stars surged in our Galaxy, with peaks at 5.7 billion years ago, 1.9 billion years ago and 1 billion years ago. These epochs correspond with our Galaxy's interactions with the dwarf galaxy. The first epoch included the time during which our Sun was formed. It seems that the birth of our Sun was triggered by the dwarf galaxy's infall.

The Sagittarius Dwarf Elliptical Galaxy is on its final pass around our Galaxy. It is being disrupted by the Large Magellanic Cloud, a galaxy of significant mass that is approaching our Galaxy and making its first fly-by. It happens that it will pass quite close to the dwarf galaxy, and the three galaxies will dance around each other. The whirl will fully dissipate the stars of the Sagittarius Dwarf Elliptical Galaxy into a continuous stream without any noticeable concentration.

Other star streams in the Field of Streams were discovered by Belokurov and Australian astronomer Daniel Zucker's team in 2006, including the Orphan Stream, whose parent has not been identified, and a trail of stars being stripped from the globular cluster Palomar 5, similar to the stream of stars left behind by Omega Centauri.

The stars that make up the streams in the Field of Streams were identified by grouping stars with similar properties in their spectra as measured by the Sloan Digital Sky Survey (see page 53), and that lined up in curved arcs in three-dimensional space in a way that suggested they were linked and enabled them to be distinguished from other stars in the Milky Way. In the course of time, over a very few orbits of the Galaxy, all the streams will become more diffuse. In time, it will become too difficult to separate the stars in this way.

Our Galaxy's companions

The Magellanic Clouds are two luminous areas in the night sky that look to the naked eye like bits broken off the Milky Way but are, in fact, two galaxies: the Large and Small Magellanic Clouds. They have been long known to the original inhabitants of the southern hemisphere, like the Aboriginal Australians, but were seen and first entered into written history by Europeans during the early voyages of discovery to the southern seas. The first drawing that survives is a star chart of 1516 made by Italian explorer Andrea Corsali, a double agent for the Medici family. On a mission to India to find commercial opportunities for the family to exploit, he reported that (in the words of a contemporary translation) 'We saw manifestly twoo clowdes of reasonable bygnesse movynge above the place of the pole, nowe rysynge nowe faulynge, so keepynge their continuall course in circular movynge.'

The clouds were named after Ferdinand Magellan (Fernão de Magalhães e Sousa), the Portuguese captain who led the first European circumnavigation of the world (1519–22). Magellan had no opportunity to tell the story of their discovery since he was killed in the Philippines during the final months of the voyage home. It was Antonio Pigafetta, an Italian navigator on the voyage, who reported that: 'The Antarctic [celestial] pole is not so covered with stars as the Arctic, for there are to be seen many small stars congregated together, which are like two clouds a little separated from one another and quite dim, in the midst of which there are one or two stars.'

The Large and Small Magellanic Clouds are the largest of the satellite galaxies of our Galaxy. The Large Magellanic Cloud is 14,000 light years in diameter and is 1 per cent of the mass of our Galaxy; the Small Magellanic Cloud is less than half that. They are 200,000 and 150,000 light years from Earth, respectively, and were once thought to be our nearest neighbours. They are definitely on our Galaxy's doorstep to intergalactic space and significantly subject to its tidal forces. The two galaxies, particularly the Small Magellanic Cloud, are the source of a stream of material orbiting and falling onto our Galaxy. The stream shows as a faint arc of neutral hydrogen gas across half the sky.

There are about sixty satellite galaxies known and nearby to our Galaxy, most of which seem to have been satellites right from the start. A few may have been captured as they passed close to our Galaxy, and some may have disappeared from view by merging with others. They range in distance from the edge of our Galaxy out to about 1 million light years. They are typically small galaxies, so-called dwarf galaxies, some only a few hundred light years in diameter or less, and containing fewer than one thousand stars. If the smaller of the satellite galaxies of our own Galaxy were within its boundaries, they could be called star clusters; what makes them galaxies in their own right is principally their isolation in space. There is thus a continuum between the satellite galaxies and globular clusters.

The satellite galaxies are a puzzle, not because there are so many but because there are many fewer than calculated. The Andromeda Galaxy has a similar number to our Galaxy; about thirty are known. The Millennium Simulation (see page 54) suggests that there should be 500–1,000 satellite galaxies in orbit around big galaxies like ours and Andromeda, and that number is about ten times the reality. The solution seems to lie somewhere in the properties of dark matter. Another puzzle concerning the dwarf galaxy satellites of our own Galaxy is that several of them contain a surprisingly large amount of dark matter.

The galaxies with which our Galaxy merged over the last few billion years were individually minor and there do not seem to be any further mergers in the offing in the immediate future. As a result, the Galaxy has kept its spiral structure intact for the past 9 billion years, settling down after each disturbance to its regular life. It will remain calm for 4.5 billion years into the future, then the Galaxy will merge with a galaxy that is bigger than ours, which will completely destroy its spiral shape, transforming the two into an elliptical galaxy, with the distinct possibility that our Sun will be thrown off into intergalactic space (see Chapter 12). A collision between our Galaxy and another triggered the birth of the Sun and a second collision will determine the circumstances of its death.

Our Galaxy's supermassive black hole ate a star and is resting after the big meal

Disturbances by passing and merging galaxies have a big influence on the black hole in our Galaxy, throwing food, in the form of gas and stars, into its gravitational maw. Our black hole was probably born as the Galaxy collapsed but its appetite has caused it to grow since then. It is classified as supermassive, although it is of a modest size compared with those in other galaxies, and is 'only' 4 million times the mass of our Sun. It was tracked down in the centre of our Galaxy early in the history of radio astronomy.

The person who discovered that the sky emits celestial radio emission was American radio engineer Karl Jansky (1905–1950), who between 1928 and 1932 worked for Bell Telephone Laboratories at Holmdel, New Jersey, investigating the sources of interference that might affect transatlantic telephony. He built an antenna in the form of an open, rectangular, wooden frame, with aerial wires strung over it. On wheels, it rotated on a track and was nicknamed 'the Merry-Go-Round'. By 1932, Jansky had found a natural source of 'static', or radio noise, that he described as 'a very steady hiss', with a maximum fixed in space along the Milky Way.

For reasons decided by his commercial employer, Jansky had to abandon his astronomical investigations in 1933, but his discovery was followed up by another American radio engineer, Grote Reber (1911–2002), who had astronomy as a hobby and built a dish-like radio telescope in Wheaton, Illinois, that was an object of curiosity for the local population. After a light plane circling the radio telescope suffered an engine failure and had to make an emergency landing, some of the locals speculated that the telescope was a weapon emitting destructive rays.

In 1939–42, Reber was the first to map the Milky Way in radio waves, and showed that its greatest intensity peaked in the constellation of Sagittarius. This radio source gathered the name

Sagittarius A as the strongest source in that constellation, or Sgr A for short. It soon became clear that Sgr A was complex, with two main halves: Sagittarius A West and Sagittarius A East. Sgr A West coincides with the highest density of stars in the Galaxy and in 1959 the International Astronomical Union agreed to make it the central node of a coordinate system to map the Galaxy as seen from our position. It was an inspired choice because in February 1974, American astronomers Bruce Balick and Robert Brown discovered a bright point-like radio source within Sgr A West. In the *Astrophysical Journal* in 1974, they concluded: 'The unusual nature of the sub-arcsecond structure and its positional coincidence with the inner 1-parsec core of the galactic nucleus strongly suggests that this structure is physically associated with the galactic center (in fact, defines the galactic center).'

The radio source became known as Sagittarius A* (pronounced 'Sagittarius A-star' and abbreviated as Sgr A*). It proved to be the black hole at the centre of our Galaxy. As it is a black hole, it is in itself invisible because no light or radio waves can escape from the strength of its gravitational field, but closely surrounding the black hole is a rotating disc of material that is falling in, like water circling the drainage hole in a bath, and radio and other emission is coming from that disc, powered by the energy released during the fall. Surrounding that is a cluster of a couple of dozen stars, extending out to a distance of some 50 light hours, about one hundred times the size of our solar system. Around that again is a dense cluster of thousands of stars extending out to a distance of several light years, comparable with, but considerably more tightly packed than, a globular cluster of stars. Intermingled with these stars are several gas clouds.

The motion of the stars surrounding the black hole has been studied by several teams, one led from the Max Planck Institute for Extraterrestrial Physics by the German astronomer Reinhard Genzel (b. 1952) and one from the University of California, Los Angeles, led by the American astronomer Andrea Ghez (b. 1965). For more than two decades, they have repeatedly imaged the inner stars with the European Southern Observatory's telescopes in Chile and the Keck Observatory twin telescopes in Hawaii and been able to see them orbit Sgr A*. The stars are speeding round Sgr A* with velocities up to 1,400 kilometres per second (3 million miles per hour). The motions arise because of the pull of Sgr A* and make it possible to estimate that it is 4.6 million times the mass of the Sun. Ghez and Genzel were awarded the Nobel Prize in Physics in 2020.

One of the stars orbiting the supermassive black hole at the centre of our Galaxy is on a particularly long, thin orbit that takes it very close to Sgr A*, within 45 astronomical units (AU), without colliding (45 AU is forty-five times the distance of the Earth from the Sun, just a bit more than the distance of Pluto from the Sun). The 4.6 million solar-mass object must be smaller than this. What can be this massive and this small? There is only one suggestion that makes sense. Calculation of the size of a black hole of that mass shows that its radius is seventeen times the radius of the Sun, so it easily fits within the star cluster and the orbit of its most closely approaching star.

The stars orbiting Sgr A* are fewer than they were. One that has left the cluster used to be in a binary system (two stars orbiting one another). Under the influence of the rest of the cluster, it ventured too close to the Galaxy's black hole. The two stars and the black hole engaged in a gravitational tussle in which the black hole, a million times more massive than either of the stars, was an inevitable victor. About 4.8 million years ago, it broke the binary star apart and threw out one of them, which became a high-velocity star, speeding through the Galaxy much faster than other stars. The way in which black holes do this was worked out in 1988 by American astronomer Jack Hills (b. 1943), then of the USA's Los Alamos National Laboratory: what happens is thus called the Hills Mechanism.

In 2019, the ejected star was identified as the one catalogued as S5-HVS1, located in the southern hemisphere below the galactic centre on the far side of the Galaxy, at a distance of 29,000 light years from Earth. It is an A-type star, which means that it has a mass of about 2.4 solar masses, and is a common sort of star, like many others – Altair, Sirius and Vega are just three of the stars in the night sky that are similar. It was picked out from others by virtue of its high speed, measured by a survey being conducted with the Anglo-Australian Telescope in New South Wales in Australia: it is moving at 1,755 kilometres per second (about 3.9 million miles per hour), speeding in a radial direction away from Sgr A*, the centre of the Milky Way Galaxy. By contrast, our Sun moves at just 225 kilometres (140 miles) per second circumferentially around the Galaxy. The trajectory of S5-HVS1 has been tracked back and it shows that the star originated from near to the central black hole of our Galaxy. Its companion is presumably still in the cluster of stars orbiting around the galactic black hole, or maybe it was swallowed by the black hole.

The velocity of S5-HVS1 is so high that it will inevitably leave the Galaxy in 100 million years and never return. It spent the first part of its life in close companionship with another star in a binary system and was separated from it about 4.8 million years ago. It is spending the middle of its life as a single star among the many stars in the crowded Galaxy and is doomed to live out the future and final part of its life in the cold darkness of extragalactic space. There will be a short period as it reaches a distance of a few million light years when it becomes first a red giant and then a beautiful but unseen planetary nebula. Beyond that time, it will die a quiet and isolated death.

S5-HVS1 is the first clear example of the Hills Mechanism in action. The star was probably thrown away from the black hole with a speed in excess of 8,000 kilometres per second (about 17.9 million miles per hour). This is nigh on 3 per cent of the speed of light, so S5-HVS1 was given quite a kick – the cosmic equivalent of being hit out of the baseball park or cricket ground to score a home run or a six. It has been slowed to its current speed by the retarding effect of the force of gravity of the black hole and the Galaxy in combination.

When a star gets very close to a black hole the force of gravity experienced by the near and far sides of the star (or cloud) can be very different – the near side is more strongly attracted to the black hole than the far side. This is a tidal force, akin to the force of the Sun that causes the tides in the sea. When the star is far away, its own force of gravity holds the two sides together, but when it is too close, the near side is lifted off the star. The tidal force distorts the shape of the star, elongating it into a thin, rope-like string of gaseous debris. This process is facetiously called 'spaghettification'.

Extreme spaghettification leads to complete disruption, and the material from the disrupted star (or cloud) can fall into the back hole, moving through the disc that surrounds it. The sudden surge of material wakes the black hole. It causes the black hole suddenly and temporarily to emit copious amounts of light, radio waves and X-rays. The technical name for such a phenomenon is a Tidal Disruption Event (TDE); the nearest and best studied TDE is AT2019qiz, which started in September 2019, came to maximum brightness in October and faded from view after five more months. The TDE took place in a black hole of mass about a million times the mass of the Sun, in a face-on spiral galaxy at a distance of 215 million light years.

The unlucky star, which ventured too close to the black hole, was similar to the Sun. It was stretched into shreds, half of which were sucked into the black hole and half churned into the space all around. This material formed an opaque wall of material around the black hole that hid the final stages of the event from view. In general, similar walls make it hard to spot TDEs. Because black holes eat stars in private, TDEs are much more frequent than our few discoveries suggest.

Estimates are that a star falls into the black hole in our Galaxy, spaghettifies and causes one of these large flares on average once every 50,000 years or so. Some encounters will have produced very bright outbursts, seen perhaps only by the uncomprehending eyes of prehistoric animals or the awed eyes of our pre-human ancestors.

A flare from our black hole

One modest outburst from our Galaxy's black hole took place three hundred years ago. From 1994 to 2005, Japanese astronomers led by Tatsuya Inui of Kyoto University collected observations made by a series of X-ray telescopes that showed how a cloud of gas near the central black hole responded to the outburst. Regions in the gas cloud called Sagittarius B2 brightened and faded over the course of nearly twelve years. The varying X-ray output from the black hole Sgr A* had taken three hundred years to travel to B2, so the cloud was responding to an event that had occurred three hundred years earlier. The brightness of the 'light echo' suggested that the black hole was a million times brighter three centuries ago than now. That would still not have made it visible to the naked eye, however – evidently the meal on which the black hole dined then was just a snack: perhaps a small comet coasting through space. The black hole would need to eat something like an entire star to make an outburst that could be seen from Earth. That meal, a full dinner, was what probably happened in the encounter of the binary star with Sgr A* 4.8 million years ago.

The most recent infalls into our Galaxy's black hole have been very small and caused only modest bursts of radio waves and X-rays. The largest seen was observed in 2013 by Chandra, NASA's X-ray sensitive telescope in space – the X-ray emission from the Galaxy's black hole increased by a factor of four hundred. There was a near miss in 2014, during an encounter between a gas cloud known as G2 and the Galaxy's black hole, which caused astronomers much excitement. It was forecast that the gas cloud would be disrupted as it approached the black hole and that, although the cloud would not hit the black hole directly, some of the material would spill into it. However, the black hole did not light up. Perhaps a massive star was embedded inside the gas cloud, with a strong enough force of gravity of its own to counteract the pull of the black hole and keep the cloud together as it passed. G2 came close to the black hole but not close enough to trigger a spectacular flare. The event was a disappointing anticlimax.

If we compare the black hole in our Galaxy with others, it is not very bright. One reason for this is that it is not nearly as massive as other supermassive black holes, which can easily be ten thousand times more massive. Another is that not much matter currently falls onto our black hole. It seems to have purged the surrounding volume of space of gas. Some matter does dribble in a thin trickle of material into the disc orbiting the black hole and then into the black hole itself. The material responsible for the radio emission from Sgr A* is thin gas emitted from stars in the vicinity, the stars in the cluster of stars surrounding the black hole. For the black hole this is not a meal, it is a between-meal snack. Apart from this minor continuous nibbling, our Galaxy's black hole is currently resting after its most recent spaghetti feast.

6

Ancestors, Siblings and Children: The Stars and the Sun

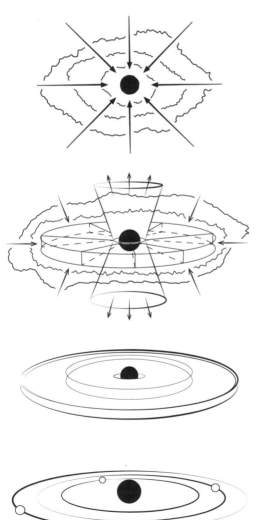

The Sun was not born from a gas cloud as a single child but as a member of a family of stars, so it had many siblings. It had ancestors, who contributed to the composition of the gas cloud from which the Sun formed. The Sun had children, the planets of the solar system. The Sun is thus connected into a family tree that has its roots back to the stars that lived during the Dark Ages and descendants that include our Earth and, one could say, ourselves.

Within the clouds of gas and dust in each galaxy, new stars condensed. They recycled some of the condensing material back into space, but some remained behind in flat discs, rotating in orbit around the stars. Material in the discs quickly aggregated into systems of planets.

The formation of our Galaxy's first stars and star clusters
Like the Galaxy itself, the stars in our Galaxy formed in collapsing clouds of gas. Each collapsing gas cloud fragmented and formed more than one star – perhaps only a few or possibly as many as hundreds or thousands of stars – all were generated at the same time, so essentially, the stars in the cluster had the same birthday. The result was an association or a cluster of stars.

The oldest star clusters in our Galaxy (other than globular clusters) are NGC 6791 and Be 17 at 10 billion years old. Star clusters contain stars of very different masses, ranging upwards from a few per cent of the mass of the Sun. It is usually the case in a collection of objects that there are more of the least massive objects and fewer of the most massive ones. On the seashore, for example, there are few rocks and boulders, a multitude of pebbles in the shingle and grains of sand that seem countless. A star cluster thus consists of a small number of large stars and a large number of small ones.

When first born in a star cluster, the larger stars are both brighter and hotter than the smaller ones, so a typical very young star cluster has a few bright, hot blue stars, a large number of white or yellow ones the same brightness and temperature as the Sun, and many faint, cooler orange or red ones. An older star cluster may have some bright red stars, which have evolved from the blue or white ones so the most massive stars have reduced in number.

Photographs of young star clusters look like a collection of gemstones, with a few bright blue sapphires, a number of white diamonds and a scattering of red garnets – one such cluster, NGC 4755, is also known as the Jewel Box, gaining its name from a remark made by nineteenth-century English astronomer John Herschel when he described its telescopic appearance as 'a superb piece of fancy jewelry'. This star cluster lies in the constellation of the Southern Cross and was discovered by French astronomer Nicolas Louis de Lacaille in 1751. It is mostly composed of blue stars but has one that is a particularly prominent bright red.

Sometimes the collapse of the gas cloud that made stars happened simply because the gas cloud grew dense and collapsed under its own weight. Sometimes the collapse may have been triggered by an external event. During an interaction between two galaxies, for example, the passing galaxy may have caused gas clouds to collide in the Galaxy, squashing them together at the interface. This can create so many stars at once that the galaxy is called a starburst (see page 75). These triggers for star formation could be the ones that occurred right from the start of the Universe, even in gas clouds made (as they were at the outset of the Universe) solely of hydrogen, helium and dark matter, and were enough to start star formation in our Galaxy some 13 billion to 12 billion years ago. Once some stars had been made, they were able to influence the formation of more stars.

As described in Chapter 4, the first stars were made of the two light elements (hydrogen and helium), which do not much impede the flow of radiation outwards to their surface from their central cores, where nuclear reactions take place. As outlined in Chapter 8, stars are a balance between the downward force of gravity and the upward push of the outflowing radiation, so the balance of these forces in the first stars is different from the balance in stars nowadays. The outcome was that, in general, the first stars were more massive than stars now and generated more energy than is usual now. The gas clouds in which the stars were embedded were strongly heated and expanded. The expanding gas was driven like a piston into surrounding gas, squashing the gas at the interface. This triggered the formation of a second generation of stars: the same repeating mechanism operates in the Galaxy today (pl. IX).

The galactic ecosystem

Because the first stars in our Galaxy were so massive and radiating so much, they evolved quickly, completing their lives in hundreds of thousands or millions of years, rather than the billions of years more typical of stars now. They turned into supernovae, exploding into their surroundings. This added further energy into the gas cloud, both in the form of radiation and the motion of the outflowing body of the star, amplifying the formation of stars in new generations.

However, the very events that triggered the formation of further generations of stars created so many that further star formation was inhibited through the process of feedback.

The energy poured into the Galaxy pushed interstellar gas out into the halo, causing it to flow up along the rotation axis of the Galaxy in the phenomenon known as the 'galactic wind'. Satellites flying astronomical telescopes sensitive to gamma rays have detected this hot gas streaming up from the galactic centre, flowing out perpendicularly to the Galaxy's disc. This outward flow initially reduced the raw material available in the disc for new stars to form. However, hydrogen gas in the outer reaches of the Galaxy, in nearby intergalactic space, was able to fall into entrances in the Galaxy, where there were open corridors around the sides. This replenished the interstellar material in our Galaxy and started a new cycle of feedback.

These cyclic processes form an ecosystem centred on our Galaxy, enveloping subordinate ecosystems within, like the cycle that connects interstellar hydrogen clouds, stars and supernovae in loops. In ecology, an ecosystem is a biological community of interacting organisms and their physical environment. Used more generally, as here, the word denotes a complex network or interconnected system. In astronomy, the Earth itself forms an ecosystem that includes its magnetic field. The solar system is an ecosystem that includes the Sun and a solar wind that extends out beyond Pluto, and planets that exchange rocky material in the form of meteoroids (debris in orbit in the solar system, broken off planets and asteroids). The stars and nebulae in the Galaxy form ecosystems that cycle gas back and forth. Galaxies are an ecosystem, with subordinate ecosystems comprising stars, gas and other components of galaxies; the physical environment is intergalactic space. Clusters of galaxies are even larger, with stars, gas and dark matter interacting and exchanging energy.

As well as the hydrogen gas that pervades the space between the stars (the interstellar medium), there is gas between the galaxies (the intergalactic medium). The cyclic process that moves hydrogen gas from the intergalactic medium to the interstellar medium in a galaxy and then to stars, and then to gas that escapes from the galaxy back into intergalactic space, is not a perfectly regenerating cycle. The basic raw material is hydrogen, made in the Big Bang, which resides in intergalactic space around the galaxy, and it is gradually being used up. Some of it escapes entirely into space, beyond the ecosystem of the galaxy. Some of the hydrogen is transformed into other elements that are carried away in the general flows, to wherever the flow goes. Some hydrogen ends up as dead stars – neutron stars, white dwarfs and black holes – which remain in the parent galaxy. Some of the mass of the hydrogen is changed into energy and is radiated away into space. However, some and perhaps most of the hydrogen that flows out of the galaxy is recycled back through the galaxy, probably many times.

We do not have direct evidence of the size of the ecosystem of our Galaxy – it is too difficult to judge the size of the ecosystem from our position inside it. However, astronomers have been able to map the ecosystem of our similar neighbour, the Andromeda Galaxy, which is about 1 million light years in radius, so it is presumed that our Galaxy's ecosystem is approximately the same size. Considering that the Andromeda Galaxy is 2.5 million light years away, it is likely that the ecosystems of our Galaxy and Andromeda's just about touch near the halfway mark. If the two galaxies were closer, they might well exchange gaseous material through a contact point. This happens in clusters where galaxies are crowded together.

Our terrestrial biological ecosystems – for example, the relationship between animals, trees, carbon dioxide and atmospheric oxygen – operate in cycles and subcycles that all lie within the galactic ecosystem. The relative size of our terrestrial ecosystems to the galactic ecosystem is the same as the relative size of a virus to the Earth. Our part in the operation of the Galaxy is, as one might have anticipated, utterly insignificant!

The ageing demography of the Galaxy

Like the history of the human population, the history of the number of stars in the Galaxy has been one of a progressive increase, while the history of their environment shows evidence of progressive change as stars have interacted with their surroundings. In human history, changes in the number of humans took place in fits and starts. The human population modified its environment for its own benefit, progressively optimizing its properties to favour human beings: the numbers of people increased. Humans also overexploited their environment and caused intermittent famine: their numbers decreased. Thus, the size of the human population has surged back and forth in cycles on top of the progressive trend. Likewise, in cosmic history the stars of the Galaxy have become more abundant over time as stars aged and died, triggering waves of new star formation. Additionally, they were born in surges, as the result of starbursts caused by successive encounters with other galaxies, such as the Sagittarius Dwarf Elliptical Galaxy (see page 105).

Stars have progressively moved interstellar gas, gathering it into their own bodies and eventually locking it away into their carcasses when they die, which has depleted the interstellar gas. The chemical composition of the gas that remains in galactic space has changed, gradually enriched or polluted (depending on your point of view) by new chemical elements created by nuclear processes in stars and in explosive events like supernovae, and spread into space by various kinds of stellar outflows and explosions. In detail, stars are nuclear furnaces, burning hydrogen in nuclear fusion processes that change hydrogen into heavier chemical elements. They include helium, carbon, oxygen, iron and all the other chemical elements. Eventually, the nuclear fuel in each star is used up and the stars die. They change into black holes, neutron stars and white dwarfs and fade away. At various stages in their lives, they blow outwards with a wind, or they explode as some form of supernova or kilonova, so that in any case they recycle some of their material back into interstellar space. The material that they eject has been enriched with heavy elements and mixes with the interstellar gas in the Galaxy. Interstellar gas was at first pristine from the Big Bang, made of hydrogen and helium only, but progressively it became enriched with new elements. New generations of stars formed in this enriched material and the cycle repeats.

As a result of this cyclic process, younger stars have larger amounts of heavy chemical elements than older ones. Astronomers use a simple, convenient shorthand for this, dividing stars into two groups (with a proposed third group: see page 124). Stellar populations are numbered in the reverse order in which they were born. Population I stars are composed of hydrogen and helium and noticeably enriched with heavier elements. Older Population II stars are composed likewise of hydrogen and helium plus far smaller amounts of heavy elements. The quantity of heavy elements is significant enough to affect the appearance and structure of the stars, and, with not very much work, the two populations are distinguishable through relatively simple procedures. The division between stellar populations correlates with the way our Galaxy and others are constructed. In a given galaxy, the stars of one population or the other congregate together. One immediately visible result of this is that the colour of a galaxy varies from place to place. In a spiral galaxy, the stars of the spiral arms are blue and the stars of its halo are red. The classification of stars in this way was developed in the 1940s and 1950s, first by German-American astronomer Walter Baade (1893–1960).

Baade was born and became an astronomer in Germany and moved to the USA and the Mount Wilson Observatory in California in 1931. He started the process to become an American citizen but in a move from one house to another he lost the papers. He disliked any bureaucratic business and failed to reactivate the procedure that would have overcome this setback. As a result, when the Second World War began, he was still a German citizen and therefore classified as an enemy alien and confined to a limited area. However, the authorities took a tolerant view and that area was Pasadena, California, which included the Mount Wilson Observatory. Like Brer Rabbit confined by Brer Fox in the briar patch, he was where he wanted to be. The other astronomers at the observatory had been drafted into the armed forces and deployed away from astronomy, so Baade had as much use of the 100-inch Hooker Telescope as he wanted. He took his work as an astronomer seriously and while observing with the telescope, he was always formally dressed in jacket and tie. Most astronomers nowadays dress to use a telescope in jeans and a T-shirt, with a warm, outdoor overjacket or hoodie, as if going for a country hike.

At 2.5 metres (100 inches), the Hooker Telescope was the largest in the world from 1917 to 1948, and Baade used it to investigate the stellar content of different types of galaxies. Wartime conditions along the seashore of California made his use of the telescope even more effective. In February 1942, a Japanese submarine lobbed artillery shells onto Santa Barbara and showed that the coastal cities were within striking distance of naval attack. Fear of air raids intensified wariness and as a result, restrictions were placed on the use of artificial lighting in the Los Angeles basin. The skies were dark, an atmospheric condition that favours the study of faint stars in distant galaxies.

With access to a large telescope under dark skies, Baade was for the first time able to discern the individual stars of M 31 (the Andromeda Galaxy) and its two companions, M 32 and NGC 205. By 1944, he had recognized the distinctive properties of the blue stars in the spiral arms of M 31, in contrast to the red giant stars in the two companions and the nucleus of M 31, and called them Population I and II respectively. By analogy, what had happened to M 31 to produce this distinction was similar to what happened to our own Galaxy, which is also a spiral galaxy.

By 1957, Baade had realized that this was not a straightforward division – there was a continuum from one population to the other – but astronomers still retain the simple idea as the basis for easy discussion. It is close enough to the history, since the underlying reason is related to the progressive origins of stars. Early stars were injected into the halo of the Galaxy by early mergers, but, later, stars formed in its disc. The stars that formed lately still include bright blue stars, and the stars from the halo have all been around for so long that they have turned into red giants and white dwarfs (see page 162). Population II stars are old and the only ones found in the halo; Population I stars are young and found in the disc, near the spiral arms of the Galaxy.

Although Population II stars in the halo have smaller amounts of heavy elements, they do have some, so they cannot have been among the very first stars. The stars made immediately after the Big Bang must have been made purely of hydrogen and helium because that was all the material available – they were the ones that made the heavy elements that we now see in Population II stars. They belong to a stage of development of our Galaxy that preceded Populations I and II. These, the earliest stars, have been given the label of Population III but none have ever been discovered. They were born a long time ago and were more massive and brighter than stars of the other two populations, meaning that they radiated prodigious amounts of energy and used up their nuclear fuel faster than is now normal, so they have all died. When astronomers get telescopes that can look back into the Dark Ages, they expect to find some Population III stars.

We humans have a stake in the classification of stars into populations. We live on a solid planet made up of heavier elements made in stars that lived previous to the birth of our Sun and we use such elements in our biochemistry. It would not have been possible for our Sun to make a planetary system with a planet like ours, nor for us to evolve, if the Sun had been born too early in the life of the Galaxy. Our Sun was born 4.55 billion years ago, two-thirds of the way through the life of our Galaxy and is a Population I star. If this had not been the case, we would not be around to discuss the issue.

What caused the birth of the Sun?

The Sun formed out of an interstellar cloud of gas when the cloud became unstable – that is, when the internal gravitational force pulling the cloud together became larger than the outward internal pressure force of the cloud. This did not happen without reason:

there was a trigger and a series of events that induced the Sun's birth. The trigger could well have been due to the infall of the Sagittarius Dwarf Elliptical Galaxy, as described on pages 105–106. Its passage through the plane of the Galaxy disturbed the interstellar gas cloud and its gas swirled about. Parts of the cloud compressed and became extra dense and then collapsed, unstoppably, forming into a star cluster.

The massive stars of the cluster evolved quickly. One of them became a supernova and caused a further collapse within a surviving gas cloud and this formed the Sun. The supernova exploded near to the gas cloud – within about 10–20 light years – and compressed its boundaries. We know this because the supernova explosion left its traces behind in meteoritic material: the Allende meteorite, for one. It is made of material preserved from the time of formation of the solar system and reveals some of its history.

The Allende meteorite was recovered for science largely through the efforts of two Smithsonian Institution scientists, Brian Mason (1917–2009) and Roy Clarke (1925–2016), and of NASA geologist Elbert King (1935–1998). King worked to prepare the programme to analyse lunar rocks brought back from the Moon by Apollo astronauts, whom he trained in geology at the Manned Spacecraft Center in Houston, Texas – he had taken up the study of meteorites as a means to develop suitable techniques.

The Allende meteorite fell 560 kilometres (350 miles) south of El Paso in Texas, in an area including the village of Pueblito de Allende, in Chihuahua, Mexico. On 8 February 1969, a bright meteor was seen in the very early morning hours. Local people described loud noises like claps of thunder as the meteorite fell, with stones falling from the sky. The meteorite was originally the size of a car but broke into hundreds of pieces as it traversed the Earth's atmosphere, scattering in a strewn field the shape of an ellipse, 50 kilometres (30 miles) long, 12 kilometres (7½ miles) wide and 250 square kilometres (100 square miles) or more in area. More than 2 tonnes were collected, the largest fragment having a mass of about 110 kilograms (240 lbs). One large fragment fell near the post office in Pueblito de Allende, missing it by only 10 metres (30 feet). Meteorites are usually named after the post office nearest to their landing site: this one emphatically selected its appropriate name.

In *Moon Trip* (1989), King described how he learnt about the Allende fall in Mexico while unsuccessfully searching for a meteorite reported to have fallen in Texas. He contacted Rubén Rocha Chávez, a newspaper editor in Hidalgo del Parral, who recounted what local people had told him about the brilliant fireball in the middle of the night and described the several pieces of meteorite he had on his desk. King immediately travelled to the city, arriving thirty-six hours after the fall. He was astonished when he saw two big meteorite pieces on the editor's desk: one weighed more than 15 kilograms (30 lbs).

King and the NASA team and the scientists from the Smithsonian obtained pieces of the Allende meteorite for scientific analysis, many of which had been collected by local inhabitants, some of whom preserved them in plastic bags previously used for food; these contaminated specimens have lost a lot of their scientific interest. The Smithsonian scientists organized schoolchildren to find pieces and instructed them on how to handle the fragments, preserving them in unused freezer bags. The youngsters were paid with bottles of soft drinks. Local people continued to find specimens, some of them large, for the next year. Small pieces are still being found and many of them are made into jewelry.

The Allende meteorite is an important and rare type of meteorite. By coincidence, it fell just as the preparations for the Apollo 11 lunar landing were nearing completion, with the prospect that the astronauts would return with lunar rocks to analyse. This focused attention on it and, with its widespread distribution among scientists, it has become the most studied meteorite ever – its properties are outlined or cited in more than 16,000 scientific papers.

The Allende meteorites contain millimetre-sized spherical chondrules (see pages 133–34) and centimetre-sized irregular mineral pieces composed primarily of calcium and aluminium oxides. These pieces proved to be 4.568 billion years old and contained magnesium-26 (Mg-26). In 1977, this remarkable discovery prompted nuclear physicists Alastair Cameron (1925–2005) and James Truran (b. 1940) to advance the idea that the nuclear processes that produced it took place in a supernova that contaminated the Sun, both stars having been born in the same gas cloud condensed by the collision with the Sagittarius Dwarf Elliptical Galaxy. The supernova was about 10 light years from the Sun. The supernova created aluminium-26 (Al-26) during the explosion and ejected it into space. Rushing outwards, the outer layers of the resulting expanding shell made from the body of the exploding star reached a dense part of the gas cloud in about a century, injecting the Al-26. This part of the gas cloud collapsed into our Sun and material left over became our solar system, including planet Earth and the meteorite. Meanwhile, because it is radioactive, the Al-26 decayed to Mg-26, with a half-life of 720,000 years. From supernova to meteorite had taken not much more than a million years. The meteorite orbited in the solar system, preserving its page of history unaltered for more than 4.5 billion years until it fell to Earth in 1969.

The rotation of the Sun

The Galaxy is rotating, but it is not a rigid object. The central parts move faster than the outer parts. At any given place in the Galaxy, one side of a gas cloud would move faster than the other. This means that gas clouds also rotate. In addition, when the gas cloud that produced the Sun was blasted by the nearby supernova, the blast would not have aligned in any way to the cloud; it would have had some glancing component. Just like a hand stroking a terrestrial globe, this would cause the gas cloud to spin. The bottom line is that the gas cloud would rotate, even if slowly.

The gas cloud was perhaps 10 light years in size. A dense fraction of the cloud, called the solar nebula, perhaps a light year or so in size, formed the Sun. The collapse of the solar nebula into the Sun caused it to rotate faster. This is an inevitable consequence of the physical principle called conservation of angular momentum. Angular momentum is essentially the size of an object times the speed of its rotation, and that product is conserved – it remains constant during a change of configuration. So, if the size of the solar nebula decreases, its rotation speed increases. A similar thing happened during the formation of our Galaxy (see pages 98–99). The Sun formed as a rapidly rotating star, potentially with a period of rotation of an hour or so. This is in contrast with the actual period of rotation as it is today, which is about one month. Just as the skirt of a spinning ice skater flares outwards due to centrifugal force, a disc of solar material spun off the equator of the Sun. Unlike the case of the ice skater, whose skirt weighs almost nothing, the disc of the solar nebula was massive, so it effectively made the solar nebula larger again. The principle of the conservation of angular momentum applies further, and this braked the Sun to its present slow rotation speed.

The nascent Sun

The supernova near to the Sun's interstellar cloud squashed the gas of the cloud in an irregular way – some parts were made denser than others, some dense regions were more massive than others. Those parts that were the right mass and the right size became unstable in a process called Jeans collapse, named after the British astrophysicist Sir James Jeans (1877–1946) who described the phenomenon in 1902. Denser-than-usual bits of the cloud collapsed where the size of the dense part was bigger than a threshold value that is nowadays referred to as the Jeans length. Different mass bits produced stars of different mass. In each case, the collapse was a runaway process, accelerating exponentially, as gravity overcame the pressure in the interstellar gas. The problem that Jeans did not solve is the same problem that urgently confronts a skier who accelerates downhill: how to stop. This part of the birth history of the Sun was figured out in 1961 by the Japanese astrophysicist Chushiro Hayashi (1920–2010).

The collapse of a small region of the parent gas cloud from which the Sun formed happened quickly; it took a total of about half a million years to form the Sun. During this time, the Sun was a protostar. The collapse built up from the centre of the proto-star, gradually encompassing the entire body of the collapsing gas cloud. The core of the protostar was made first. Material from the collapsing gas cloud's outer regions followed in afterwards, falling on the core, building its mass. Gravitational energy released in the collapse heated the protostar, which emitted submillimetre and long-wavelength infrared radiation. Clouds of dust and gas condensed onto and around the proto-Sun, turbulent and moving rapidly as they rained down. Some of the material that did not build onto the proto-Sun gradually settled into a rotational motion in a disc in a flat plane. This was the proto-solar system. The disc would eventually become the orbital plane for the Sun's planets.

Disc material flowed inwards from the outside edges towards the central proto-Sun. Colliding at the centre of the disc, material that was not accreted by the proto-Sun was spewed outwards in a jet, spraying up into the poles of the disc. There may have been residual gaseous material above the disc that was impacted by the jet. If there was, the jet material blew into the gas and excited it to emit infrared and visible radiation. If we could look back into the past and see the proto-Sun and proto-solar system at this stage, about 100,000 years after collapse began, we would see an infra-red-emitting star, surrounded by a dark disc, with a jet coming out and, a short distance away at the end of the jet, a Herbig-Haro (or HH) object.

HH objects are nebulae that were initially discovered by American astronomer Sherburne Wesley Burnham. Their importance was uncovered by two astronomers working at first independently and then in tandem, the American George Herbig (1920–2013) and the Mexican Guillermo Haro (1913–1988). Herbig had become dedicated to astronomy at the early age of eight years old, and was appointed straight from his astronomy studies at the University of California to the Lick Observatory on Mount Hamilton in California, working there until he moved to Hawaii just before his retirement. By contrast, Haro studied law and started his working life as a reporter, becoming interested in astronomy only when he interviewed the director of the Tonantzintla Observatory in Mexico. He rapidly advanced and was appointed to succeed as director in 1950. Herbig became interested in infrared-emitting stars and the nebulae with which they were often associated in the sky. He met Haro at a conference in 1949 and learnt that he was also interested in the same things. Working together, they elucidated properties of HH objects, discovering that as nebulae they were unusual in being excited not by hot stars but by collisions with other fast-moving gas. This gas, it was later discovered, had been emitted in a jet by a nearby protostar.

After about half a million years had elapsed, the proto-Sun was approaching a temperature of about 4,000 degrees Celsius (7,000 degrees Fahrenheit). It became visible to the outside Universe by blowing away the surrounding cocoon of material raining down on its surface. It was rotating quickly, and it was embedded in a dusty, gaseous environment, with everything moving quickly.

Our Sun at this stage was a T Tauri star, which is a type of variable star. Many stars of a similar sort are known today in interstellar clouds. The archetype star, T Tauri itself, was discovered by English astronomer John Russell Hind (1823–1895). While still a teenager, Hind started his astronomical career as a 'computer', performing grinding mathematical calculations at the Royal Observatory in Greenwich under the Astronomer Royal, the fearsome George Airy. He escaped this drudgery aged twenty-one, when he was employed in 1844 by George Bishop, a rich wine merchant, who had a private observatory in Regent's Park, London. Hind began a search for the planet that was perturbing the motion of Uranus. His technique was to compare the sky with star charts of the ecliptic regions, which he had prepared at the start of his employment. He was beaten to the discovery of the planet Neptune in 1846 by the German astronomer Johann Gottfried Galle. He repurposed his search towards 'minor planets', as he called them, or asteroids as they are now known. At the time, they were regarded as scarcely less interesting than Neptune. He went on to discover eleven asteroids, as well as a number of comets.

Hind had already discovered six asteroids when, on the night of 11 October 1852, he was scanning his telescope across the constellation of Taurus, near to the Pleiades and Hyades star clusters. He noticed a star in the sky that was missing from his chart – the implication to Hind was that he had succeeded in finding a seventh minor planet, one that had moved onto the area of the chart and was only masquerading as a star. The right explanation was that the object was in fact a star, one that was variable in brightness. It had been too faint to be seen when the chart was compiled but had become bright enough to be visible by the time that Hind looked. It was identified in the constellation with the letter T.

T Tauri and its surroundings became additionally attractive for astronomers when Hind noticed a small nebula not far away, which disappeared a decade later. It returned to view briefly in the 1890s and came back fully in the 1930s. It was catalogued as NGC 1555 but is more familiarly known as Hind's Variable Nebula and it reflects the light of T Tauri. Not only is the star T Tauri variable so that the nebula varies in brightness too, but there are also opaque clouds streaming in space between the two that create a play of shadows on the nebula.

In T Tauri and its nebula we have a model for the nascent Sun at its first shining about a million years after it began to collapse from its parent gas cloud. It emerged from its opaque envelope of stellar formation. Having recently coalesced from its dusty and gaseous surroundings, it became visible at optical wavelengths. Like T Tauri itself, the Sun would have been changing rapidly in brightness, due to violent activity in the atmosphere of the star and to moving clouds of dust and gas.

In the proto-Sun, the downward force of gravity became progressively opposed by an upwards push caused by the increasing outward flow of energy from within. As the proto-Sun got hotter, it became transparent to the flow of radiation. It began to generate nuclear energy, increasing the amount that it radiated even more, so that the downward pull of gravity was exactly balanced by an upward push by the out-flowing radiation. At this point, the proto-Sun settled down into the Sun, having achieved the status of a mature star.

From the beginning of the collapse of the proto-Sun, triggered by the impact of material flung out by the nearby supernova, until the protostar became the Sun took about a million years. During this time, the solar nebula developed its disc-like shape, with the Sun at its centre, generating energy and warming the inner parts of the nebula.

Composition of the solar nebula

On 28 September 1969, just before eleven o'clock, a meteor broke up with a bang over Murchison, in Victoria, Australia. It startled the cows in a dairy farm and attracted the attention of people going to church that Sunday morning. Because it arrived at such a crucial time, there were many eye witnesses to the meteor, which broke into meteorites that fell across the town. (Rocks that orbit in space are meteoroids, rocks that then traverse the Earth's atmosphere are meteors, and rocks that reach the ground are meteorites. The same rock may change its designation three times in a matter of seconds!)

Brothers Peter and Kim Gillick, aged ten and eleven, were building a ferret cage in their backyard when the meteor exploded over their heads; the explosion set off an intense interest in them that lasted over a year: they became meteorite hunters. One meteorite punctured the roof of a hay barn: it had obviously fallen from the sky and became the model for what they were looking for. It was very distinctive – black, crumbly and smelling of methylated spirits.

The boys plotted maps of the pieces that were accidentally recovered. They realized quickly that the small pieces travel a shorter distance and fell quicker to the Earth, while the larger pieces, less influenced by air resistance, carried on and went further. They extrapolated from their map to identify further areas worth searching. The meteor was travelling towards a lake called the Waranga Basin and the biggest piece could well still be under water, but over the next twelve months, the boys recovered from dry land about one-third of the more than 80 kilograms (180 lbs) of the meteorite that has been gathered. The family donated some of what they had found to Australian institutes and museums and sold most of the rest, making enough to pay for the boys' college education.

Fragments had been promptly collected and carefully preserved in clean, plastic bags, so they were not contaminated, and ideal for scientific analysis. They showed that, like the Allende meteorite, the Murchison meteorite is a carbonaceous chondrite.

Meteorites are pieces that originated as asteroids, solar system bodies that used to be called minor planets, a name that expresses accurately what they are. Like the other planets, minor planets formed from the solar nebula. Some asteroids are large, rounded worlds that settled under their own gravity. Such a large asteroid like this trapped heat generated internally from radioactive elements and generated externally by the fall of smaller asteroids onto its surface. It melted inside. Minerals that became liquid at low temperatures, such as iron, percolated into a central core, leaving more rocky material wrapped around the asteroid in a mantle near the surface. The asteroid 'differentiated', or separated into zones (see Chapter 10). If such an asteroid breaks up (because it collides with another one), the pieces are made of iron or rocks depending on what part of the asteroid they come from.

Lots of meteorites are like this, but not the Murchison meteorite. It originated from another kind of asteroid, a small world that never became massive enough to become spherical. The insides of asteroids like this never melted, so the material of which they are made never differentiated. The material of the solar nebula packed together enough so that it fused into a solid but never suffered from a strong pressure or a high temperature, although it may have reached a temperature and pressure high enough to alter its mineral composition. (Exposure to water is another factor that might have changed its mineral composition.) Small asteroids may never have collided at a high speed with other asteroids. Such a small asteroid may, however, now encounter the Earth and fall as a meteorite. It will have a distinctive appearance – individual small pieces of rock, or chondrules, millimetres in size, fused together.

A chondrite is such a meteorite: stony with chondrules, the small spheres of silicate rock.

Within the class of chondrites there are a number of different sorts, and carbonaceous chondrites, like the Murchison meteorite, are of a type that contains abundant carbon and organic compounds. The carbon compounds make the meteorite black. The remarkable smell of the Murchison meteorite was so strong that it was noticeable at the time of the fall, diffusing down from the meteor's trajectory, as well as diffusing from the pieces themselves. The smell comes from organic molecules, like amino acids, sugars and alcohol-related chemicals, in the material of which the meteorite is made. These organic molecules may be the seeds from which life formed on Earth, brought to our planet by a similar meteorite in the distant past. The molecules had not been generated by living organisms, but it could be that on Earth they associated together and produced biology-like structures that became live creatures.

The thought is that the molecules had originally been made in the solar nebula (see Chapter 9) – the same material as the gas cloud that was the birthplace of the Sun. The cloud was primarily hydrogen gas as made from the Big Bang, with a few per cent of helium, but it was enriched with elements like carbon, oxygen, silica and iron made in stars and exploded into the interstellar medium, including elements made from the nearby supernova, like magnesium-26, made from the radioactive decay of aluminium-26. It also contained elements formed by dying stars billions of years ago. As stars age, the nuclear processes inside them make elements such as carbon, oxygen, silicon and calcium. The stars also change structure, and the changes dredge up these heavier elements from internal regions to their surface. They blow these elements out into the surrounding environment, where they cool and coalesce into molecules. These molecules condensed into solid grains in the solar nebula as the Sun was forming, and then into the planets and other objects in the solar system that were forming at that time, including the Allende and the Murchison meteorites. As mentioned on page 126, the oldest grains are 4.568 billion years old, as determined from the properties of the radioactive elements that they contain, and this is what is taken as the age of the Sun and the solar system.

Presolar grains are small solid pieces in the solar nebula that predate the formation of the Sun and the Murchison meteorite is the most abundant source of such grains that we know. A similar dust grain, technically a calcium-aluminium-rich inclusion (CAI), has also been found in a fragment of the Allende meteorite. The inclusion was named 'Curious Marie' by University of Chicago scientists who, in 2016, found the element curium (named after Marie Curie) in it; the name is both a pun and a further tribute to the notable woman scientist and Nobel Prize winner. In 2020, scientists led by Olga Pravdivtseva, working at Washington University in St. Louis, determined that the CAI predates the formation of the solar system.

Presolar grains in both the Murchison and Allende meteorites came from other stars. They formed in the outer parts of a dying star and got pushed away by the radiation pressure from the star into the surrounding environment. They became part of the interstellar medium and travelled in the Galaxy. In this way, the gas cloud from which the Sun formed became contaminated with dust that had its origin in other stars that pre-existed the Sun. This dust became incorporated into the solar nebula and then found its way into the planets and the fragments of planets that we call meteorites.

The dust of the solar nebula that accumulated into the Earth has lost its identity over time, crushed and remoulded by geological and biological activity. It has been remixed into new chemicals and new structures, such as mountains, trees and people: the composition of the material of your own body can be traced back to the solar nebula. The connection between the material of your body and the solar nebula is, of course, a complicated one, made up of many links, and it would be difficult to draw conclusions about the solar nebula from human biology. But there are bodies from the solar nebula that have not undergone such radical processes and are much more primitive: they are chondrites like the Allende and Murchison meteorites. In the construction of a chondrite the dust of the solar nebula has been squashed into a solid, only gently and without much modification. Chondrites are the closest we have to the original material of the solar system.

What happened to the Sun's family?

The Sun was born 4.55 billion years ago – it has since left home, lost contact with its kin and grown to maturity.

It is not known for sure that the Sun was formed as part of a compact cluster of perhaps as many as a thousand stars, about 10 light years in dimension, but that is a scenario that is often discussed. In any case, few if any stars are formed in isolation. The Sun is thus very likely to have been formed as part of a family, a village or even a city of stars.

Virtually all the stars in the Sun's family would have formed with a planetary system. The stars were originally very close, each separated from its neighbour by perhaps only 1 light year on average, and sometimes passing much closer. In those encounters there was the opportunity for the planets of one star to cross into the planetary systems of the other. The planetary systems of each star may now consist of its own planets and planets captured from elsewhere.

In fact, this may be the case for our own solar system. Most of the planets of the solar system appear to have been formed in the same process that formed the Sun, although perhaps some of the asteroids are adopted. Among them are a few that move in tilted orbits among the outer planets: these asteroids are collectively known as Centaurs. They are a mixed bag, but the orbits of some of them have been tracked back to their birth orbits well beyond the furthest members of our solar system and lying across the plane of orbit of the rest of the planets. They may have been pulled into our solar system from another planetary system around one or more of the Sun's close siblings in its birth-cluster.

Although born into a close-knit family, the Sun and its solar system are now living on their own. The closest star is called Proxima (meaning 'the nearest'), which is 4.22 light years away from the Sun, equivalent to 40 million million kilometres (25 million million miles) – the Sun's nearest neighbour is actually quite far away – and lies close to a bright star, Alpha Centauri, which is a multiple star. The brightest star, Alpha Centauri A, has a companion, known as Alpha Centauri B. They have a 'high, common proper motion': namely, they move across the sky, they move very quickly and they move together. These are signs that A and B are joined together in a binary star system and that they are not far away – in general, the closer something is, the faster it seems to change position. If you lie on your back in a meadow, the bees flying over your nose zip across your field of view, whereas a high-flying jet aircraft seems to crawl across the sky. The distance of the two stars is very small as star distances go: at 'only' 4.37 light years, or 41 million million kilometres (26 million million miles) from the Earth.

In the course of a study of the Alpha Centauri binary star, the Scottish-born astronomer Robert Innes (1861–1933), working in South Africa, discovered a rather faint star that had the same motion as the other two, a fellow traveller and apparently part of the Alpha Centauri system, which it indeed proved to be. For Innes, this was a discovery towards which he had been working for most of his life and might explain why, at the time, he rather over-sold the significance of what he had found.

Innes went to school in Dublin, where he showed an aptitude for mathematics, but as one of twelve children he was obliged to leave school at an early age. He worked in commerce but remained an autodidact. In 1890, he moved with his new bride to Sydney where he became a successful wine merchant, which meant that he was able to indulge an interest in astronomy, in particular the discovery and measurement of binary stars and other mathematical problems. In 1896, because of his business success he was offered a position as an administrator at the Royal Observatory in South Africa, with the promise of some involvement in its scientific work. Some of it was paid but he carried out much of it unpaid, in his spare time, making a success of both aspects of this post.

In 1903 he was appointed the founding director of the newly established Union Observatory in Johannesburg and began a survey of the Southern sky looking in the area around high proper-motion stars to find similar ones. His aim was to find nearby binary and triple stars and, in 1915, he discovered the third star in the Alpha Centauri system, the companion to A and B; initially, it was given the neutral name of C. Innes and Dutch astronomer Joan Voûte (1879–1963) independently measured the distance of C, showing that A, B and C are indeed all members of the same triple star system. A and B are close to one another, orbiting with a period of eighty years. C moves around the pair of them in a very elliptical orbit with a period of half a million years and at a considerable distance from the two, currently, 13,000 times the Earth–Sun distance.

Voûte published his data on the distance of C before Innes. Innes's data suggested that C was not only close to Alpha Centauri A and B, but also closer to Earth than either, so he rushed into print soon after Voûte with that claim. His results did not really justify this assertion, because, like all data, it was uncertain and, although his claim was possibly correct, the uncertainty was large enough that it was also more or less equally possible that it was not. Without a close analysis of the accuracy of his data, Innes took the chance, named the star Proxima and was acclaimed for its discovery. More accurate measurements over the years have shown that his guess was, in fact, right and Proxima became recognized as the nearest star to Earth, not counting the Sun.

'Near' is, of course, a word to be appreciated in its astronomical context. Travelling at 299,800 kilometres (186,000 miles) per second, the light of Proxima takes 4.24 years to travel to us. The triple system is on an approaching trajectory and in 27,000 years, Proxima will make its closest approach to the Earth at only 3.0 light years. At about the same time, it will lose the title of 'nearest star' because its orbit will have taken it to the far side of Alpha Centauri A and B, which will then alternate as the nearest star to the Sun, as the one orbits around the other with their period of eighty years.

The Alpha Centauri system also contains three planets (a, b and c) that orbit Proxima in its planetary system. Proxima b is much like the Earth in size and mass, although considerably closer to its parent star than Earth. Proxima is rather a dim star but because Proxima b is relatively close to it, it also has about the same temperature as the Earth. The surface environment of the planet is thus quite Earth-like, making it an interesting target for investigations of astrobiology, the science of life in the Universe. Can life exist there – even alien intelligence? Given the planet's closeness to us, could we communicate with extraterrestrials? One's imagination can readily spiral off into optimistic predictions!

To come back closer to reality, does Alpha Centauri C deserve the name Proxima – is it in fact the nearest star to the Sun or could there be an even closer star, as yet undiscovered? From time to time, astronomers have sought to explain some feature of the solar system by the effect of a star nearer than Proxima. For example, in 1982, University of Chicago palaeontologists Jack Sepkoski and David Raup studied mass extinctions – global catastrophes that wiped out many species at the same time – as they appeared in the geological record of fossil strata and noted an apparent periodicity in the times of the extinctions that indicated some recurrent event. Working at about the same time, American physicist Adrian Melott and palaeontologist Richard Bambach homed in on the periodicity as 26 million years during the last 500 million years, later modifying their analysis of the periodicity to 60 million years.

In 1984, astronomers Daniel Whitmire and Al Jackson, and independently Marc Davis, Piet Hut and Richard Muller, sought to explain the extinctions as the result of meteoric and cometary bombardments of the Earth, like the Chicxulub asteroid that struck Mexico 64 million years ago and caused the extinction of the dinosaurs (see Chapter 11). They formed the hypothesis that the bombardments are triggered by periodic disturbances of asteroids in the solar system, by an undiscovered star, which, they further hypothesized, is in a periodic highly elliptical orbit. It disturbs the solar system at its closest approaches to the Sun, a position known as perihelion. If the periodicity was 26 million years, the star would lie at a distance of only 1.5 light years at the present time. The star was named Nemesis, after the goddess who controls fate; it is also popularly known as the Death Star, following the nomenclature of the movie series, *Star Wars*.

To be close but remain undiscovered, Nemesis must be a faint, cool, brown dwarf star. Brown dwarf stars emit infrared radiation. There have been several general surveys for infrared-emitting stars, and some surveys targeted specifically towards finding Nemesis. No star as near as 1.5 light years has been detected. Its existence, and indeed the existence of a periodicity in the occurrence of mass extinctions in the prehistoric history of life on Earth, remain speculative, and even unlikely.

So, as far as we know, the Sun is alone. What has happened to the Sun's birth companions?

It is sometimes possible to identify 'moving groups' of stars orbiting the Galaxy on parallel tracks. A moving group began as a cluster of stars and the stars have 'remembered' the original trajectory of the cluster. Astronomer Marina Kounkel and her colleagues at Western Washington University in the USA have used Gaia space satellite measurements of the movements of over 1 billion stars to uncover nearly 2,000 moving groups of stars in the Galaxy, up to about 3,000 light years from us. Around half of these stars are found in long, string-like configurations that move as a group.

The moving groups that survive in our Galaxy are composed of young stars because after just a few million years the stars lose their ties with their original family. The reason is that, over time, the stars in the cluster drive out all the residual gas that remained after their formation and weaken the force of gravity that tethers the stars to their cluster of origin. This happened long ago to the Sun and its siblings. They have journeyed around the Galaxy about twenty times in their lifetime, not only having lost the gas whose mass kept them tied together, but also being jostled by random encounters with other stars and with gas clouds. As a result, the stars of the solar cluster have gradually separated. The Sun's siblings have mingled into the crowd of other stars of the Milky Way, lost their family connection and become anonymous.

7

The Sun: A Star in Its Maturity

The Sun is the source of all the energy that we consume as humans (except for nuclear power, which comes from the Big Bang or from long-dead stars that exploded as supernovae). Its importance to us would alone justify the Sun having a chapter to itself, but in addition it is an utterly typical star and, with the rest, is crucial to the biographical development of the Universe.

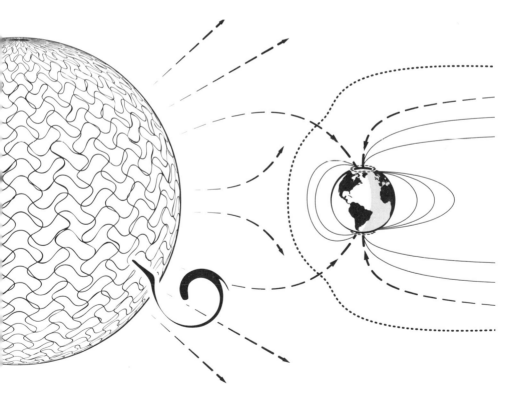

The Sun and its wind of energetic particles dominates the space surrounding the planets, including the Earth. The Earth's surface is, however, shielded from the Sun's particles by a magnetic field.

The origin of the Sun's energy

While the Sun was a youngster, a protostar, it radiated energy from a range of sources such as the gravitational energy that was dissipated by its fall from a cloud in the interstellar medium into a star, and for some time in the nineteenth century astronomers thought that this might still be the source of the power of the Sun: the German physicist Hermann von Helmholtz and the Canadian astronomer Simon Newcomb calculated that the Sun could have existed for perhaps several millions of years if it were so, a period of time thought to be indescribably long. However, in the twentieth century, scientists began to realize that the Earth was much older even than this, and since the Earth was dependent on the Sun, the Sun must be as old or older than the Earth.

The twentieth-century conclusion about the age of the Earth was based on the new science of radioactivity, discovered by French scientists Henri Becquerel, Marie Curie and Pierre Curie. It enabled the British physicist Lord Ernest Rutherford to develop a way to use radioactive decay to measure the age of rocks. A young American chemist, Bertram B. Boltwood, discovered that some rocks were as much as 1 billion to 2 billion years old, and we now know of rocks two to four times older. This was far too long for the gravitational energy of the Sun's formation to have lasted. How could the Sun keep shining for such a long time?

The answer was discovered by two physicists at the University of Göttingen in Germany – Fritz Houtermans (1903–1966) and Robert d'Escourt Atkinson (1898–1982). Holidaying together on a walking tour in the summer of 1927, they talked about the source of the Sun's power. Atkinson, a British astrophysicist, knew that Sir Arthur Stanley Eddington (see page 17) had just determined the physical conditions inside the Sun and how it maintained its size: its density and its temperature created a high pressure inside the Sun, which countered the force of gravity that was drawing the material of its body tightly together. The Sun is in balance between these upwards and downwards forces respectively. The balance is very precise because the pressure and temperature are self-adjusting. If the Sun expands for some reason unconnected with its own structure, the pressure and temperature will fall and it will contract. The opposite is also the case. It always acts to restore its equilibrium.

The pressure comes from two sources. The gas of which the Sun is composed generates its own pressure, but in addition, the radiation generated in the core flows up through the Sun. Both pressures push against gravity and support the Sun against collapse.

Atkinson knew that the high density and temperature inside the Sun would mean that what was then called atomic transmutation of the solar material was possible. The atoms (or, as we now know, their broken-down nuclei) in the centre of the Sun would frequently collide. If the collisions transformed some atoms from one kind to another, losing mass in the process, what was then called atomic energy (we would now more precisely call it nuclear energy) would be produced. 'This might be the source of the Sun's energy,' suggested Atkinson. 'Let's just work the thing out, shall we?' said Houtermans, a nuclear physicist. 'How could it happen in the Sun?' It did not take the two men long to work it out on a piece of paper. Houtermans was able to boast to his future wife about the sparkling stars that the two of them saw during an evening walk, telling her: '"I've known since yesterday why it is they sparkle." She didn't seem the least moved by this statement. Perhaps she didn't believe it. At that moment, probably, she felt no interest in the matter whatever.'

The two young scientists had discovered how the fusion of light elements into heavier ones could fuel the Sun. In these reactions, four hydrogen nuclei become fused into one helium nucleus. The mass of the helium nucleus is less than four hydrogen nuclei, and the loss of mass is converted into energy, through Albert Einstein's famous equation of the equivalence of mass and energy, $E = mc^2$. The amount of material that the Sun fuses in its core is enormous: 620 million tonnes (683 million short tons) of hydrogen into 616 million tonnes (679 million short tons) of helium each second. Thus, each second, the Sun converts more than 4 million tonnes

(4.4 million short tons) of matter into energy. The Sun is so large that this attrition rate can be kept up for billions of years. The Sun was a protostar for only half a million years. It will spend a total of about 10 billion years fusing hydrogen to helium and has been 4.5 billion years doing this so far.

What happens inside the Sun?

The details of what happens inside the Sun have been confirmed with astonishing accuracy by the detection of neutrinos (tiny particles given out in the nuclear processes), which travel from the Sun's interior and have been detected on Earth using specially built neutrino detectors.

When we see the Sun through mist or as it sets in a dusty atmosphere, we can see that it has a surface. In scientific terms this means that sunlight originates from a narrow layer in the Sun – the Sun has a transparent atmosphere above that layer through which light can travel, but below that layer the Sun is opaque. As a result, we cannot look at the interior of the Sun directly. At first only the Sun's surface characteristics and its global properties – such as its diameter and the amount of energy that it radiates – could be determined by direct observation. Its interior was literally out of sight.

However, we now know what happens inside the Sun. Ingenious mathematical calculations have built up a theoretical picture of the Sun's interior in such a way that the results of the calculations fitted the Sun's global properties. The understanding of the inner workings of the Sun that developed was the result of one of the great feats of reasoning in modern physics. From the 1920s, astronomers knew the physical conditions inside the Sun by calculation and from the 1930s they knew that nuclear reactions were the source of the Sun's energy. In the 1950s, they had begun to understand the way that stars evolve in relation to one another from observations of star clusters. These calculations built up confidence in astronomers' theoretical knowledge of the Sun's interior.

This knowledge was further developed by tuning the calculations to fit what could be learnt from two radiations that pass from its interior regions through solar material and that provided clues as to the conditions in the Sun's interior regions. The radiations that opened the inner workings of the Sun to direct scrutiny are neutrinos and sound.

Neutrinos are made in the proton-proton nuclear chain reaction inside the Sun that makes the Sun's energy. Because it is so hot in the Sun, the hydrogen atoms there split apart and become free electrons and protons. Because it is so dense in the Sun, the protons readily hit one another. Two of them may combine, one of them changing to a neutron by emitting a neutrino and a particle called a positron. The reaction continues towards its conclusion when another proton sticks to the pair, forming a helium nucleus containing a pair of protons and a neutron. Two similar helium nuclei then collide and two protons are ejected, leaving behind a helium nucleus with a pair of protons and a pair of neutrons. The net result of this chain is that four protons (p) have made a helium nucleus (^{4}He), releasing energy, two positively charged positrons (e^+) and two neutrinos (v_e – there are three kinds of neutrinos and this is the kind associated with electrons). In the notation of nuclear physics:

$$4p \rightarrow He + 2e^+ + 2v_e$$

The neutrinos escape at the speed of light, travelling so fast that it takes only eight minutes for them to reach the Earth. They carry small parcels of energy. The numbers of neutrinos given off by the Sun is immense – 65 billion of them pass through every square centimetre (0.2 square inch) of the Earth every second. Reckoning your cross-section at about 1 square metre (10 square feet), that is 100,000 billion passing through you every second.

There are floods of solar neutrinos, but they are whisper-quiet, few interact with anything and we have no sensation from them. A typical neutrino can travel through a light year (10 million million kilometres/6 million million miles) of material without interacting with it in any detectable way. Not many interact with the thousands of kilometres across the diameter of the Earth, let alone the few centimetres of flesh on a person.

Despite the astonishing speed and elusiveness of neutrinos, it is possible to build detectors that do catch some solar neutrinos because they are not entirely inert and there are so many that a tiny number do interact. The first solar-neutrino detector was built by American physicist Raymond Davis, Jr (1914–2006), of the Brookhaven National Laboratory in the USA, following technical suggestions from the Italian-born physicist Bruno Pontecorvo (1913–1993) and American physicist Luis Walter Alvarez (1911–1988).

Davis worked with American astrophysicist John Bahcall (1934–2005), who insisted that it was practical to try to catch neutrinos, as the vast numbers constantly released by the Sun overwhelmed the small chance for each one that it would be caught by a neutrino detector. In the bowels of the Homestake Gold Mine, in Lead, South Dakota, deep enough underground to avoid interference from cosmic rays, Davis installed a tank containing 615 tonnes (677 short tons) of carbon tetrachloride, a solvent normally used for dry cleaning. (The amount was equivalent to 380,000 litres/100,000 US gallons – the volume of water contained in a large swimming pool. The physicists joked that, if they were unsuccessful in their experiment and lost their jobs, they could set up in the cleaning business.)

Solar neutrinos were captured on the chlorine atoms in the solution and converted to radioactive argon atoms. These argon atoms were flushed out of the tank every few months and counted as they decayed by emitting a radioactive particle. (This method of counting the argon atoms was why the detector was put underground: the number of radioactive particles from cosmic rays that could get muddled up with the argon atoms had to be kept to a minimum, absorbed by 1,478 metres/4,850 feet of rock above the experiment.)

Bahcall originally estimated that Davis would capture just seventeen argon atoms from the tank in each extraction run, but in fact, in the first experiment in 1968, lasting six months, many fewer neutrinos were seen, about one-third as many as calculated. The calculations were scrutinized and the equipment was improved, but the same result appeared when the experiment was re-run. The question was 'where are the missing neutrinos?', which became known as the 'solar neutrino problem'. Davis's experiment could detect neutrinos but not say where they came from.

Another neutrino detector called Kamiokande was built and operated by Japanese astrophysicist Masatoshi Koshiba (1926–2020). Kamiokande was able to determine the trajectory of the incoming neutrinos, and, looking back along their paths, was able to prove that the neutrinos the detector captured came from the Sun. It thus confirmed in 1989 that Davis had indeed detected neutrinos from the Sun and that there were fewer than expected.

The Sudbury Neutrino Observatory (SNO) was located 2,100 metres (6,800 feet) underground in Vale's Creighton Mine in Ontario, Canada. It detected solar neutrinos through their interactions not with carbon tetrachloride but with a large tank of 'heavy water'. Heavy water is water made, as ordinary water, of an oxygen atom and two hydrogen atoms, but the hydrogen atoms each consist of one electron in orbit around a nucleus that is made, not of a single proton, but of a proton and a neutron. Such a hydrogen atom is called deuterium, for which the symbol is D. The chemical formula for water is H_2O; heavy water is D_2O.

Heavy water is used as a moderator to control nuclear reactors. It exists in nature but is rare – in naturally occurring water, about 1 molecule in 20 million is heavy water. It is correspondingly expensive to separate the heavy water from the others. To detect neutrinos, SNO used about 300 million Canadian dollars' worth of heavy water, from a stockpile kept by Atomic Energy of Canada with which to make its CANDU (Canada Deuterium Uranium) reactors. The heavy water is not used up in the process, so SNO was only borrowing the water and gave it back when the experiment was ended in 2006. SNO likewise found that solar neutrinos were missing.

At first, some physicists thought that the discrepancy between observation and theory had arisen because astronomers' standard calculations relating to the solar interior must be flawed. There were no missing neutrinos, they thought, because somehow astronomers had overestimated the numbers of neutrinos that the Sun was making. The astronomers rejected this, in part because they had found a second way to look inside the Sun, to check their theories about its make-up.

This approach was called helioseismology – the study of oscillations in the body of the Sun, which resemble earthquakes studied by seismologists on Earth. In the general turmoil of motion of hot material in the Sun's interior, the Sun generates sound waves whose resonances travel across the body of the Sun. Its surface oscillates up and down. The Sun rings, like a cymbal quietly singing as it is brushed by a succession of impacts from a stream of sand grains. Of course, sound cannot travel through space. It travels through the material of the Sun to its surface, where astronomers measure the surface moving, just as the trembler of an electric bell oscillates.

The first solar oscillations were discovered in 1960 by Caltech physicist Robert Leighton (1919–1997) with the Mount Wilson Observatory's 60-Foot Solar Tower Telescope. He measured the main oscillation period at about five minutes, but there are other periods in the data that depend on the route that the sound waves take through the body of the Sun and the time that this takes. In the 1970s, physicist Roger Ulrich at the University of California, Los Angeles, suggested that the duration, frequency and tone of these oscillations could provide clues to the composition of the Sun's interior. The different zones in the Sun vary in composition, temperature and density and these conditions affect the time it takes sound to cross the Sun. The sound waves thus carry information about the interior of the Sun to the surface, just as the oscillations of earthquakes at the surface of the Earth carry information about its interior structure – this is the way that geologists know about the existence and properties of the Earth's liquid iron core, for example.

Although solar oscillations were discovered by an Earth-based telescope, these telescopes are limited in what they can find out about them. An individual telescope cannot observe the Sun after it disappears daily below the horizon at night and that limits the accuracy and the completeness with which astronomers can enumerate the sound frequencies in the data. Astronomers therefore set up networks of ground-based solar telescopes around the world to follow the Sun continuously. The networks have names like GONG (the US National Solar Observatory's Global Oscillation Network Group), BiSON (Birmingham Solar Oscillations Network) and HiDHN (High Degree Helioscismology Network) – but technical issues and bad weather still interfered with the observations.

The Solar and Heliospheric Observatory (SOHO) satellite, a joint project involving ESA and NASA, avoided this limitation. It has been staring at the Sun continuously from space since its launch in 1995. The mission was intended to last two years but has already lasted twenty-five, and will likely operate for more. The comprehensive observations of the SOHO satellite provided new data on the temperature inside the Sun, and the way that its interior rotates slower than its surface layers, generating a hot layer inside the Sun that is the ultimate cause of features on its surface. SOHO also proved that the standard calculations that had been used to measure sound-speed at various depths in the interior of the Sun were 99.9 per cent accurate. The conclusion was that astronomers knew rather well how many neutrinos the Sun was making, and Davis's 'missing neutrinos' were not the result of a miscalculation of solar conditions.

Assuming that astrophysicists knew about the state of material inside the Sun and nuclear physicists knew how many neutrinos that would create, physicists had to concentrate on the issue of why many went missing from Davis's neutrino detector. Something evidently happens to neutrinos after they leave the Sun. The explanation was first proposed by the famous physicist Bruno Pontecorvo (who had also become notorious outside his field by defecting to join the Soviet nuclear programme) only a year after Davis first found the solar neutrino discrepancy in 1968.

Neutrinos come in three different kinds – or 'flavours' – each associated with another kind of particle: electron neutrinos, muon neutrinos and tau neutrinos. They can oscillate from one 'flavour' to another as they travel the eight-minute journey across the distance between the Sun and the Earth. Davis's neutrino detector had the capacity to capture and detect solar neutrinos of only one flavour – the flavour generated deep within the Sun. By the time the neutrinos arrived on Earth, many of them had changed by 'oscillating' from that flavour to another, so they bypassed the detectors and went missing. The evidence that this happens was discovered by the Japanese Kamiokande detector and the Canadian SNO and became ever more convincing between 1998 and 2001, and beyond.

Astronomers were proud that their meticulous work on the Sun had led to a new discovery about particle physics: neutrino oscillations. The work was justly recognized by the award of the Nobel Prize in Physics in 2002 to Masatoshi Koshiba and Raymond Davis 'for pioneering contributions to astrophysics, in particular for the detection of cosmic neutrinos'. Canadian astrophysicist Arthur B. McDonald, the director of the SNO experiment, was likewise awarded a share of the Nobel Prize in Physics in 2015 for the experiment's contribution to the discovery of neutrino oscillation.

The faint young Sun problem

Calculations about the Sun have stood up well against the intense scrutiny of helioseismology provoked by the solar neutrino problem, so that its structure is well established at this point in its lifetime. But, of course, the Sun cannot have the same structure all its life – it is using up its fuel and radiating energy, and its interior changes accordingly. Among other things, the luminosity of the Sun changes.

The Sun is gradually becoming hotter, because the helium atoms that are made in the core occupy less volume than the hydrogen atoms that were fused. The core is therefore shrinking. Because the layers of the Sun become closer to the centre, they experience a stronger gravitational force. To maintain the balance between the internal pressure of the Sun and its downwards gravitational force, the internal pressure increases, which raises the internal temperature and pressure, which increases the rate at which the nuclear fusions occur, with greater power output. As a result, the Sun is 20–30 per cent brighter now than it was 4 billion years ago.

The internal structure of the Sun at the present time is so well established by results from solar neutrinos that it could almost be called a fact rather than a theory. In a way, this is not surprising because the theory is a combination of precise physics. It is a consequence of the way the Sun generates energy by nuclear fusion of hydrogen to helium in its core, and the balance of its gravitational force against the pressure of the gases inside. The calculations of all these processes are very robust. As a result, few people doubt the calculations of how the luminosity of the Sun has changed: it was fainter in the past. Putting this conclusion about the solar physics against what is known about the Earth's climate in that early era creates what is known as the 'faint young Sun problem'. For 2 billion years the young Sun was so faint it did not radiate the warmth to keep the Earth from freezing over. And yet, from about 4 billion years ago, the Earth was covered in oceans (see pages 222–23): there was no one alive at that time, but if there had been, people who looked at its surface would have seen storms of rain lashing onto the wavy surface of blue oceans, not blizzards of snow falling onto glistening white ice. The Earth should have been in an intense ice age, but it was not.

How can astronomers and geologists be so sure about this? The solar part of the faint young Sun problem seems to be in good shape. There is considerable doubt, however, about how the Earth's climate changes, even over decades, as the current arguments show about the response of the weather to changes in carbon dioxide levels generated by anthropogenic emissions. What evidence is there about the climate from so long ago?

Some of the Earth's oldest rocks are sedimentary rocks that are part of a banded iron formation and are embedded in greensand rocks from southwest Greenland at a place called Isua. Zircons in the greensand rocks are dated by radioactive techniques at 3.8 billion years old. Some of the minerals of the Isua banded iron formation can only form under surface water. The rocks were deposited in a similar process to the way that limestone is deposited from seawater in marine environments of the present time.

The Isua greenstone belt also contains pillow lavas, which are lavas shaped in metre-sized lumps that look like pillows; the structure forms as hot lava flows into cold water and occurs in modern times at places such as underwater volcanoes. The fact that they occur in the Isua greenstone belt shows that there were large lakes or oceans on the Earth's surface at least 3.8 billion years ago. Thus, there are signs in ancient rocks that the Earth was wet 4 billion years ago, in spite of the faintness of the Sun. Something was compensating for the lack of warmth from our star. This must have had an effect on the start of life on Earth, although it is hard to say what.

Most studies suggest that the 'something' was the extra blanket of an enhanced greenhouse effect. At this early time in the Earth's history, before plant life evolved to inject oxygen into the Earth's atmosphere, gases like ammonia and methane, as well as carbon dioxide, were more common and could well have contributed to a stronger greenhouse effect than is the case now (see Chapter 11). Or perhaps, although the total radiation from the Sun was fainter than now, some key wavelength regions were stronger and this affected the transparency of the Earth's atmosphere in some crucial way.

Other suggestions are based on the possibility that the Earth has changed its orientation in space and its rotation rate. In modern times, Earth rotates once every twenty-four hours around its axis, which is tilted at 23.5 degrees away from the pole of the ecliptic (Earth's orbital plane). Neither a change of the axial tilt (obliquity) nor of the rotation period directly affects the average energy received from the Sun, but they can, in principle, change the distribution of energy over the Earth. This affects the extent and distribution of ice cover. High obliquity has been shown to lead to a warmer climate and could offset the faint early Sun for axial tilt values of 65–70 degrees. What we know from records of the magnetic field of the Earth suggests that its tilt has been remarkably low and stable over the last 2.5 billion years. This evidence does not take us back as far as 4 billion years, but theory suggests that the Moon stabilizes the obliquity, so it may have always been about the value that it is now.

Earth's rotation period is known to have changed considerably as the result of tidal friction, which causes the Moon to move further away from Earth and the Earth's rotation to slow down over time. At the time that the Moon was formed by the impact of a protoplanet onto the embryonic Earth (see Chapter 10), its rotation period was about five hours. Its rotation period 4 billion years ago has been estimated to be just fourteen hours. A rapid rotation rate increases the temperature difference between equator and poles because it changes the mid-latitude eddies in the atmosphere: at faster rotation rates, these eddies are smaller and thus less efficient in transporting heat polewards. This effect could, in principle, prevent low-latitude glaciation.

The faint young Sun problem has been known for half a century and after decades of research it 'refuses to go away' (climatologist Georg Feulner, quoting geoscientist James Kasting in 2010). Since it does not seem to be a problem with the astronomical calculations, it looks as if the solution must contain something significant about the Earth. At the moment the book about the Earth's history is closed at these pages so we cannot easily read what is written there. It is unsettling that the evolution of the Earth, and the life on it, depended on something so mysterious.

The Sun's domain

The centre of the Sun is at a density of 150 grams per cubic centimetre (87 ounces per cubic inch) and a temperature of about 16 million degrees Celsius (29 million degrees Fahrenheit). The nuclear reactions that occur in the core are not very powerful – about the same per cubic metre as a cold-blooded animal or a compost heap – but the Sun is huge. It contains a lot of cubic metres and its immense power derives from its size. The density and the temperature of the Sun's core drop sharply out to about 25 per cent of the Sun's radius. The majority of the Sun's bulk and its material consists of hot plasma (gas made of ionized atoms) and radiation that flows up and out from the hot core to the cooler surface.

Although individual photons of light move at, of course, the speed of light, the photons ping-pong back and forth in the material, making little net progress outwards. The tide of outflowing solar energy thus moves slowly, taking about 30 million years to traverse the Sun from centre to surface, even though the Sun's radius is only 2.3 light seconds long (the technical name for this way of measuring the journey time is the Kelvin-Helmholtz timescale). So not only do we see the Sun's surface in the past as it was eight minutes ago (the time it takes light to traverse the 150 million kilometres/93 million miles from the Sun to Earth), but also we sense through its warmth the Sun's interior as it was in the even more distant past, 30 million years ago, because of the Kelvin-Helmholtz timescale.

Above about 70 per cent of the Sun's radius, its material circulates up and down, to and from the surface in columns, like cumulus clouds that mark convection columns in the Earth's atmosphere. Seen from an aircraft flying high above the Earth, the tops of the clouds form a mottled pattern hiding the ground below. Likewise, the tops of the solar convection columns are visible at the Sun's surface in mottled patterns known as granulation.

Effects from the circulation of the solar material generate a magnetic field, which pervades the Sun, extending out into the surrounding solar system. It pushes up through the granulation, in places bending the columns and shouldering them aside. These areas manifest themselves as sunspots, dark areas between the bright tops of the granulation. Sunspots usually come in pairs, one of them where the magnetic field exits through the surface and one where it goes back in. The dynamo that generates the Sun's magnetic field strengthens and weakens, with an eleven-year period. The abundance of sunspots follows the same cycle, with many large sunspots at sunspot maximum (the time of maximum magnetic strength) and few or none at all at sunspot minimum.

The sunspot cycle is not completely regular. From 1645 to 1715, the Sun completely missed three solar cycles, a period known as the Maunder Minimum, when very few sunspots were seen. This coincided with the phenomenon known as the Little Ice Age, when

Europe was plunged into unusually cold winters, a time of frozen rivers and deep snow. The Sun is known to affect the climate – for example, the price of wheat mimics the eleven-year sunspot cycle, a correlation interpreted as a link from the price of wheat to the harvest size to the weather to solar activity. However, the coincidental occurrence of the Little Ice Age and the Maunder Minimum might be just that: a coincidence.

The term 'Maunder Minimum' comes from the British astronomer Edward Maunder (1851–1928), who first published papers on the subject in 1890 and 1894. Although he did not in fact do the work that he described, the name Maunder Minimum is still apposite because his wife, astronomer Annie Maunder (1868–1947), was responsible for the work and made the discovery. It was another of those cases in which a female scientific collaborator had her thunder stolen. Annie had been educated in mathematics at Girton College, Cambridge, and persisted in seeking a job at the Royal Observatory, Greenwich, as a 'lady computer' in the solar department, a post below her capabilities. She married Maunder, who worked on solar topics at the observatory. As required by the conventions of the day, she was obliged to resign her job on marriage, but continued to research in solar astronomy, using family resources, and to publish through or with her husband.

The temperature of the surface of the Sun is 6,000 degrees Celsius (11,000 degrees Fahrenheit). The atmosphere of the Sun is somewhat cooler a little way above the surface but then gets progressively hotter, reaching astonishing temperatures measured in millions of degrees in the region known as the corona. The name is a reference to the crown-like halo that surrounds the Sun when the bright light of the surface is covered by the Moon at a total solar eclipse. The Sun's halo is scattered sunlight much like the Earth's blue sky, but additionally there are emissions from strongly fractured atoms that betray the corona's high temperatures. One such previously unidentified emission, known since 1869, proved in 1940 to arise from iron atoms from which thirteen of their twenty-six electrons had been ripped. Such an energetic state for iron was so unusual that it was more credible at the time to identify the emission as coming from a previously unknown element, which was named 'coronium', but that turned out not to exist. The way the Sun heats its corona and generates such mysteries remains itself a mystery.

Solar flares: 'space weather' and the Carrington Event

The corona is threaded by the Sun's magnetic field. There are arcs and bridges that arch from one sunspot to another, and holes and pillars that extend into space (pl. x). Luminous clouds of plasma zoom along the magnetic field lines. Like tangled elastic, the magnetic field lines can work their way into a tight knot and then suddenly release, creating a solar flare. Associated with flares are catapulting clouds of plasma that burst off the Sun into space, as if the Sun had sneezed. These are so-called coronal mass ejections. They may travel towards the Earth and envelop it in a cloud of electrons, accompanied by electrical discharges and luminous displays. Events like these affect the Earth's magnetic environment and create disturbances known as geomagnetic storms, or more colloquially, 'space weather'. Small events take place frequently, but they become more and less frequent in a cycle that follows the sunspot cycle. Flares are generally more powerful at sunspot maximum.

Solar flares are labelled in classes with letters: A, B, C, M, X, and then on in the series X1, X2, X3 and so on, doubling up in power each time. X-class flares occur at a rate of about ten per year, with X10-class flares occurring a few times per solar cycle. The largest solar flare recorded since satellites started to measure them scientifically in 1976 was an X28 solar flare that occurred on 4 November 2003.

The largest geomagnetic storm ever recorded took place in September 1859, after a bright flare on the Sun. It was the first solar flare ever observed, as well as the most powerful. The 1859 flare is estimated to have been X50, a 'super X-class flare', 4 million times more powerful than the flare of 2003. It is remarkable that it is known to human history because it affected only a small area of the Sun, lasted only minutes and there was no equipment with which to see it better than with the human eye. Fortunately, it was seen by an amateur but very knowledgeable English astronomer, Richard Carrington (1826–1875), who spent many hours looking at the Sun through his telescope and was rewarded by his unique discovery.

The son of the owner of a profitable brewery, Carrington was at first destined for the Church and entered the University of Cambridge to study theology, but his aptitude for science and mathematics took him into astronomy, and, on graduating, he worked in the observatory at Durham University. He fell out with his supervisor there (his quarrelsome manner made similar events a recurrent feature of his life) and, with financial support from his father and the brewery (which he had to manage), he moved to Redhill in Surrey where he built an observatory. He used his very well-made and expensive telescope to measure star positions, but also became interested in the Sun, noting that previous observers had not brought the same accuracy of measurement to observing sunspots as was routine for observing stars.

With a colleague, Carrington organized a day/night schedule to carry out his solar and stellar observing programmes and set himself the target to monitor sunspots for the whole of the eleven-year sunspot cycle starting in 1853. That programme came to a premature end because his method of observing the Sun with the unaided eye became obsolete, he became ill and a series of consuming and dramatic events in his personal life led to a tragic death. But before this decline set in, on 1 September 1859, Carrington finished his morning observations of the Sun, and then noticed two intensely bright patches in a large group of sunspots. In 1860, in an article in the journal of the Royal Astronomical Society, he wrote:

> I…noted down the time by the chronometer, and seeing the outburst to be very rapidly on the increase, and being somewhat flurried by the surprise, I hastily ran to call someone

to witness the exhibition with me, and on returning within 60 seconds, was mortified to find that it was already much changed and enfeebled.

The bright flare had lasted for only five minutes. Carrington had been unable to find quickly anyone from his household to confirm what he was seeing, but another amateur astronomer, Richard Hodgson, was also looking at the Sun from his own observatory at the same time and, in the article following Carrington's, said that he had recorded on the Sun's surface 'a very brilliant star of light...dazzling'.

Visiting the Kew Observatory in Richmond to see if they had recorded the flare, Carrington was disappointed that they had no record of the Sun for that day. Kew Observatory, however, was not only an astronomical observatory, it also observed the Earth's atmosphere. From 1845, it had recorded the atmosphere's meteorological and electrical properties with automatic equipment; the apparatus incorporated some of the earliest successful scientific photographic cameras. The observatory also observed the Earth's magnetic field. Its magnetometers were needles hung on silk threads, pointing along magnetic field lines, showing the direction to magnetic north and the angle by which the field lines dipped down into the ground. Lights shone onto the needles caused tracks on rotating drums wrapped in light-sensitive paper. Carrington was shown the magnetometer records – a few hours after the solar flare, there was an unusual crotchet-like kink in the track, showing that there had been a disturbance of the Earth's magnetic field. This coincidence gave him the thought that the Sun affected the geomagnetic field.

Additional effects of the 1859 Carrington Event – which we now know was indeed geomagnetic as well as solar – were that telegraph systems in Europe and North America failed, some giving their operators an electric shock. Over the following nights, bright displays of the aurora were seen from all over the world including Australia, Belgium, Bermuda, Britain, China, Colombia, Cuba, France, Hawaii, Japan, Mexico, Norway and the USA. The aurora over the Rocky Mountains was bright enough to imitate dawn and wake gold miners, who began preparing breakfast. People could read newspapers by the auroral light.

Solar flares (pl. x) and the associated coronal mass ejections can produce electrical and magnetic disturbances of the terrestrial magnetic field. Additionally, high energy radiation from the flares disturbs the balance of electrons, protons and atoms in the layer of charged particles in the atmosphere that lies between 80 and 1,000 kilometres (50 and 620 miles) high and which constitutes the ionosphere. The effects interrupt shortwave communications, cause electrical surges in power lines, produce auroral displays, interfere with GPS navigational signals and permanently damage electronic equipment in space satellites.

The flare of 2003 was one of a series recorded as a stuttering out of the Sun over a three-week period. The SOHO satellite, the Advanced Composition Explorer, many satellite TV and radio satellite services and a number of US Department of Defense satellites were among the spacecraft whose operation was interrupted. A Japanese satellite, ADEOS-2, was damaged beyond repair, as was one of the instruments aboard NASA's Mars Odyssey mission. The crew of the International Space Station, Commander Mike Foale and Flight Engineer Alexander Kaleri, had to shelter from the high radiation levels in the more robust parts of the spacecraft, the aft end of the Russian-built Zvezda Service Module. Airlines and ground controllers experienced communications problems with aircraft flying polar routes and diverted aircraft to safer routes and lower altitudes, using the atmosphere to reduce radiation exposure for passengers and crew but running up fuel costs and losing payload capacity. An electricity blackout occurred in southern Sweden as a result of the solar activity.

According to a report by insurers Lloyd's of London and Atmospheric and Environmental Research, a consultancy in the United States, if the Carrington Event recurred and the subsequent coronal mass ejection did envelope the Earth, it would cost billions of dollars and full recovery from the damage would take four years. This is the economic fear behind the drive by several spacefaring nations, including the USA and the UK to name just two, to forecast space weather, just like meteorological weather. If this can be done, we may be able to take some mitigating actions to avert service interruption and equipment damage. It will not be possible to stop the Sun from sneezing, but we may be able to learn enough to forecast when it is likely to do so and to turn our head so we do not catch cold.

8

Stars that Die: The Biggest Bangs since the Big Bang

Size matters very much for a star, governing its internal make-up and life expectancy. As a star's fuel runs out, its energy production weakens and its pressure and temperature adjust. This leads eventually to the collapse of the star into a white dwarf, a neutron star or a black hole — or even to its complete destruction.

All stars originate in interstellar clouds, and much of their body cycles back to interstellar space when they die. Massive stars evolve and explode as supernovae, making black holes and neutron stars (right loop); less massive ones, like the Sun, make planetary nebulae and white dwarfs (left loop).

Dying stars

The way that a star dies differs from star to star principally because of the star's size, or, more accurately, its mass. For humans, there is room to debate whether, as Neil Young sang in some versions of 'Hey Hey, My My (Into the Black)', 'it's better to burn out than to fade away'. Stars go one way or the other, according to their parameters (but have no opinion about which is better). Likewise, the age at which a star dies also differs from star to star. The more massive a star, the more nuclear fuel it has, so the natural expectation would be that massive stars live longer, but that is not, in fact, the case. The more massive a star, the more internal pressure it requires to support itself, and, while it is living, the star adjusts its temperature and density to provide a high pressure. The rate at which nuclear reaction occurs inside the star depends very strongly on temperature and density, so massive stars are very luminous.

In general, the less massive a star is, the longer it lives. Stars that are less than about 0.9 times the mass of the Sun can live for longer than the Universe has existed, so all of them that have ever been born are still alive (unless something unusual and catastrophic has happened). Stars between 0.9 and 8 solar masses last for many millions or even billions of years. Stars at 8 solar masses last for 30 million years. More massive stars live for even less time than that.

White dwarfs and Type Ia supernovae

A star like the Sun with a mass up to 8 solar masses turns into a red giant star towards the end of its life, burning the helium that it has produced in its core. The helium is transformed to carbon and oxygen. This phase of its life lasts perhaps a billion years. The red giant star sheds its outer layers as a so-called planetary nebula. The residual core of the star then quickly turns into a white dwarf star, with a mass typically of 0.6 solar masses but up to about 1.4 solar masses (the mass of the most massive white dwarf star so far discovered is 1.3 solar masses). Nuclear burning has ceased in such a star, which shines only by radiating the heat that it still contains from this earlier nuclear burning. It progressively cools, turning eventually into an all but invisible black dwarf.

White dwarfs take their time to transform into black dwarfs – probably none in our Galaxy have ever made it to this stage yet. The dimmest white dwarfs known have a cooling age of 9 billion years. They typically have come from stars originally of about 3 solar masses, which took about 300 million years to become white dwarfs. Looking on the oldest white dwarfs as having come from the oldest stars in the Galaxy, this puts the age of the Galaxy as 9.3 billion years, at a minimum.

There are billions of white dwarfs in our Galaxy: 95 per cent of stars end their lives in this way. However, although they are common, they are faint, so they are easily overlooked and none had been discovered until 1910. The first to be identified is called 40 Eridani B – the letter 'B' refers to the fact that it is the faint companion to the star 40 Eridani A, a binary star discovered by William Herschel (and further shown to be a triple).

Because it is in a binary system, it is possible to infer the mass of 40 Eridani B, which is not unusual and not very different to the Sun, at 0.6 times the Sun's mass. On a routine visit to Harvard College Observatory, American astronomer Henry Norris Russell (1877–1957) from Princeton University pointed out to Observatory director Edward Pickering (see page 41) that 40 Eridani B was abnormally faint. They discussed the consequence that it must be rather small – small stars have less surface area from which to radiate light, do not radiate much and therefore are faint – 40 Eridani B would thus seem to be abnormally dense, which implied that the structure of the star was different from others. Russell mentioned that it would be useful to know the star's temperature so that its size could be properly determined – the amount of light per unit area radiated by a star depends on its temperature, so dividing its total luminosity by the light radiated per unit area leads to an estimate of the star's surface area and hence a more precise determination of its radius. The Harvard Observatory was in the middle of a project to find the temperatures of large numbers of stars. Pickering made

a telephone call to his assistant, Scottish astronomer Williamina
Fleming (1857–1911). Russell recalled that 'in half an hour she
came up and said "I've got it here...". I knew enough, even then,
to know what it meant.'

The temperature of the star was very high: it was 'white' hot.
But it was also very dim, which meant that it was very small – a
'dwarf'. Russell correctly surmised that the star was a similar size
to the Earth, much smaller than the Sun or other similar stars
although its mass was like theirs. The term 'white dwarf' for such
a star was coined by Dutch-American astronomer Willem Luyten
in 1922.

White dwarfs are extremely small and dense – a matchbox filled
with white dwarf material would weigh 1 tonne – and the force of
gravity at their surface is very strong. White dwarf material is extraor-
dinarily incompressible since it has to withstand the tendency of
the star to collapse under its own weight. In 1925, a young British
physicist, Ralph Fowler (1889–1944), using the new science of
quantum mechanics, discovered that this material is 'degenerate':
all the electrons in the material are packed together as closely as
is physically possible, stopped from getting closer by the laws of
quantum mechanics. The pressure generated in the degenerate
material resists the tendency of the star to collapse, even given the
strong gravitational force that drags it down.

Fowler's discovery was incorporated into calculations of the
structure of white dwarfs by a nineteen-year-old Indian mathematician Subrahmanyan Chandrasekhar (1910–1995). In 1930, he was
on an ocean liner sailing from India to Britain to study at Trinity
College, Cambridge, passing the time on the cruise around the
Cape of Good Hope by making relaxing astrophysical calculations
(unlike most other passengers, one may safely guess). Chandrasekhar
discovered the relationship between the mass and the size of white
dwarfs – he found to his surprise that the more massive the star,
the smaller its size. In fact, he found that there is a mass at which
a white dwarf would be point-like. Above this mass, white dwarfs
cannot exist, no matter how massive a star that they formed from.

This limit is known as the Chandrasekhar mass and is about one and a half times the mass of the Sun.

A white dwarf star at the Chandrasekhar limit occupies just a point – an infinitely small volume, or in the language of mathematics, a 'singularity'. If nature nears a mathematical singularity, it is heading towards a physical impossibility. Before the impossible is reached, nature diverts into some other realm. The new realm here is an explosion and, possibly, the creation of a black hole. Not everyone realized this at first. When Chandrasekhar presented his results to his colleagues in 1935, he was publicly ridiculed by the most distinguished astronomer in Britain at the time, Sir Arthur Stanley Eddington (see page 17), who called the result 'stellar buffoonery'. Humiliated, Chandrasekhar abandoned his career in Britain and emigrated to the USA, where he worked at the University of Chicago for the rest of his life. His lifetime of achievement was recognized by the award of the Nobel Prize in Physics in 1983 'for his theoretical studies of the physical processes of importance to the structure and evolution of the stars'.

Rarely, a white dwarf is one member of a binary star system so compact that the white dwarf catches any material that floats off its companion. It gets more massive, and, in conformity with Chandrasekhar's calculations, it gets smaller. It may accrete so much that eventually it collapses towards a small point. This scenario was proposed in 1973 by the American theoretician Icko Iben and the young British astronomer John Whelan to explain some extremely bright stellar explosions, those called Type Ia supernovae. The white dwarf's collapse releases huge amounts of energy and triggers a thermonuclear explosion that destroys the white dwarf star.

An alternative idea that leads to the same result is that two white dwarfs might exist in a binary star. They might touch and merge, and head towards being a white dwarf star that is over the Chandrasekhar limit and too massive. This, too, might produce a Type Ia supernova.

The most famous Type Ia supernova was the first supernova observed scientifically. It was monitored by the extraordinary Danish astronomer Tyge Ottesen Brahe de Knudstrup, better known as Tycho Brahe (1546–1601). Tycho was a rich nobleman, who indulged his appetites and his interest in astronomy by applying both his own wealth and that of his king, Frederick II. He was notably eccentric and had a gold artificial nose, having lost part of his actual nose in a duel. He had a pet elk that died after falling, drunk, down a flight of stairs. After a change of royal rule in Denmark, Tycho lost financial support and spent his final years in Prague under the patronage of Emperor Rudolf. After a formal dinner there, he died after medical complications brought on from retention of urine, having been too embarrassed to excuse himself from the table. His monument in the Prague church in which he is buried shows him in armour, with a large moustache, his self-indulgence evident from his obesity and his jowls.

Tycho had taken up astronomy as a teenaged student and became obsessed by discrepancies in tables in his textbooks that gave the positions of planets and stars. With meticulous care he made sighting instruments that were the best possible at the time – pivoted arrangements of wooden rulers with sights and brass protractors to measure angles – and he set out to improve the data by making the most accurate measurements that he could. One evening in 1572, presumably after dinner, Tycho was driving home in his carriage and saw a group of peasants marvelling at something in the constellation of Cassiopeia. What he saw where they were pointing was a new bright star in the night sky. Tycho recorded his discovery in his book *De Nova Stella* (About the New Star), boasting about his familiarity with the stars and giving the original observers little credit in his story.

Where credit is certainly his due, it is that for over a year from his home at Herrevad Abbey, then in Denmark, Tycho repeatedly measured the position of the new star (now known as Tycho's supernova) and proved that it did not shift its position in the slightest relative to other stars in the constellation. The significance of this was that, if a star was close by, it would shift its position as the Earth rotated and took Tycho from one side of the planet to the other.

The star had to be at a distance well beyond the orbit of the Moon: it had, in fact, to be a star, not a meteorological phenomenon, as was thought at the time about any celestial object that changed (for example, had been invisible and then appeared).

Tycho and others also made assiduous measurements of the brightness of the star from its sudden appearance on about 1 November 1572 until it faded below naked-eye visibility in the spring of 1574. The observations that were made compared the brightness of the new star with other fixed stars, and it has been possible to use these comparisons to show accurately how the star's brightness changed. The plot of the changes has a characteristic shape, the same as modern light curves of Type Ia supernovae.

Although it has long faded from view, the spectrum of Tycho's supernova was obtained in 2008 by the National Astronomical Observatory of Japan's 8.2-metre (320-inch) Subaru Telescope on Mauna Kea, Hawaii. The flash of light from the supernova is being reflected from a screen of dust in the Milky Way off to the side of the direct journey from the supernova to us, taking an extra 436 years to travel the greater distance of the diverted journey. The reflection had been discovered after a targeted search by the 4-metre (150-inch) Mayall Telescope at the Kitt Peak National Observatory in Arizona. So, as well as the light curve, the spectrum also proved that the type of Tycho's supernova was due to a thermonuclear explosion of a white dwarf star.

In 2004, the probable truth of Iben and Whelan's scenario was proved by a team led by Spanish astronomer Pilar Ruiz-Lapuente, who discovered the companion star released by Tycho's supernova. Ruiz-Lapuente used the 4.2-metre (165-inch) William Herschel Telescope on La Palma in the Canary Islands to identify the star, Tycho G, confirming her discovery with the Hubble Space Telescope. The companion star to the white dwarf, the one that donated the extra material, had been released at high speed like a stone from a slingshot, as if the white dwarf holding it in orbit had disappeared – which effectively it had. The companion star was a rather ordinary star, similar to our Sun, identified as of interest because it was moving at the expected high speed away from the position of the supernova.

Neutron stars, pulsars and Type II supernovae

The Swiss-American astronomer Fritz Zwicky (see page 26) and his German-American colleague Walter Baade (see page 122) coined the term 'supernova' in 1934. The Latin word *nova*, meaning 'new', had been used historically in astronomy for a new star, and Zwicky and Baade did indeed see bright stars appear in other galaxies suddenly and without warning, as if new. For a few days, they outshone the light from all the rest of the billions of the stars in the galaxy put together. They were thus highly energetic outbursts, which merited adding the prefix 'super' to make the word 'supernova' to describe what the two astronomers had seen. The new stars faded away as the explosions dissipated.

As described above, Type Ia supernovae are collapses of white dwarfs, the corpses of stars like the Sun that lived and died in previous generations. There are other kinds of supernovae – Type II – that are collapses of the core inside a living, massive star. By far, most stars end their lives as white dwarfs, but not the rarer, more massive ones that are more than about eight times the mass of the Sun.

If a larger star with too massive a core tries to create a white dwarf, it cannot – the putative white dwarf star implodes. The energy that the collapse releases blows off the outer layers of the star in a fierce explosion. These layers contain chemical elements that have been generated by the nuclear processes that have been powering the star, so the ejected material contains large amounts of helium, carbon, oxygen and other elements that were in the star, as well as the elements manufactured in the explosion itself, like iron and nickel. This material forms a large shell around the site of the explosion, which collides with interstellar material all around. The collision heats the ejecta and the free electrons in space to make it a source of not only light but also X-rays and radio waves – a 'supernova remnant'. It remains visible in space for thousands of years until it dissipates.

Supernovae occur once every few decades in a galaxy like ours, but we miss seeing most of them – they occur in stars that are behind opaque curtains of interstellar dust. There are historical records of six supernovae in our Galaxy in the last thousand years. The supernova of 1054 was noted by Chinese astronomers and depicted on the Bayeux Tapestry, which records the invasion of Britain in 1066 by Duke William of Normandy. This supernova produced not only an outrushing nebula, called the Crab Nebula (pl. xi), but also an energetic star, the stellar remnant of the supernova and a pulsar, which was discovered in 1968 by American astronomers David Staelin and Edward Reifenstein at the National Radio Astronomy Observatory in Green Bank, West Virginia.

The term 'pulsar' is a blended neologism from the phrase 'PULSating radio stAR', which describes a star that emits pulses of radio waves. The first of these stars were identified with a radio telescope in 1967 by British astrophysicist Jocelyn Bell (b. 1943) during her PhD project at the University of Cambridge. She was being supervised by Antony Hewish. Hewish, but not Bell, was awarded the Nobel Prize in Physics in 1974 in a story of scientific recognition that is seen as reminiscent of the case of Annie Maunder (see page 155). The astonishingly regular radio pulses of pulsars seemed artificial, so, although they were manifestly in interstellar space, Bell's colleagues half-seriously contemplated whether they were communication devices that extraterrestrial space travellers had placed as an Interstellar Positioning System, an IPS like our GPS. In fact, they proved to be tiny, rotating stars, spinning rapidly, sometimes in much less than a second, beaming out radio waves that sweep across the Earth on each rotation. The Crab Nebula pulsar rotates thirty times each second.

The stars that form the basis for pulsars are somewhat similar to white dwarfs but are one hundred times smaller, perhaps 12 kilometres (7½ miles) in radius – the size of a ring road around a city, not, as are white dwarfs, the size of the entire Earth. Pulsars are made of neutrons and are a type of neutron star, which are produced by the collapse of the core of stars that are between about eight and perhaps thirty times the mass of the Sun. Like white dwarfs, neutron stars have a mass comparable to the Sun. As previously noted, a matchbox full of white dwarf material weighs 1 tonne; a similar quantity of neutron star material weighs 1 million tonnes.

Tiny neutron stars are somewhat difficult to find in the vastness of space, but they help to call attention to themselves by shouting 'Look at me!' in different ways: pulsars radiate radio waves; other neutron stars radiate X-rays. If a star is in orbit around a neutron star, the star may, as it ages, swell up to a giant and leak. Some gas will fall onto the neutron star and will be compressed by a strong force of gravity. Just as air compressed in a bicycle pump gets hot, so too does this gas get hot – to temperatures in excess of a million degrees, emitting X-rays. Scorpius X-1, the first stellar X-ray source discovered in the constellation of Scorpio in 1967 by Italian-American astrophysicist Riccardo Giacconi, who was awarded the Nobel Prize in Physics in 2002, is a neutron star in such a binary star system.

Stellar black holes

White dwarfs are stars that are the basis of Type Ia supernovae. Neutron stars are created from Type II supernovae, but they are not the only possible product. A star more massive than those that formed neutron stars and whose core collapses as a supernova produces a black hole. Black holes originating in this way have masses measured in solar masses, not millions or billions of solar masses, so, in order to distinguish them from supermassive black holes (see page 66), they are termed 'stellar black holes'.

Isolated stellar black holes are dark stars that drift invisibly in interstellar space. They show themselves if matter (gas) falls into them from something nearby. In such a case, the gas releases gravitational energy, which heats the gas, perhaps to millions or tens of millions of degrees. This can happen if the black hole has a companion star that leaks gas onto it: the black hole becomes visible as an X-ray binary star.

X-ray binary stars are pairs of stars, each orbiting around the other, so close that their orbital period is short (minutes to days). One component is a more-or-less normal star. The other is a star much smaller in size (but not in mass) like a black hole. The normal star transfers matter (gas) onto its compact companion via an accretion disc: matter spirals inwards and falls onto the black hole. The matter is squashed as it falls into the throat of the black hole and gets hot. The first such black hole was discovered in 1971 in the X-ray source Cygnus X-1; about twenty are now known.

Gravitational wave events: binary neutron star and black hole mergers

When a binary star revolves, the gravity around it changes, sending ripples through space at the speed of light. These ripples are called gravitational waves and are a feature of Einstein's general theory of relativity (see page 16). They are emitted from anything in which the distribution of mass changes. They are very weak, even from something the mass of a star, and were first detected in 2015 by two exquisitely sensitive instruments, operated together as the Laser Interferometer Gravitational-Wave Observatory (LIGO). There were, however, earlier signs that gravitational waves exist, convincing enough to determine the specification of the instruments, develop their technology and gather the finance to make them.

The effects of gravitational waves were discovered even as long ago as the late 1970s, from studies of the orbits of binary neutron stars. The first of them, B1913+16, was discovered by American physicist Russell Hulse (b. 1950) and his PhD supervisor American astrophysicist Joseph Taylor (b. 1941) in 1973 with the Arecibo Observatory's 305-metre (1,000-foot) diameter radio telescope in Puerto Rico, which, alas, collapsed in 2020 following damaging hurricanes. Its significance for general relativity was immediately realized. It was a Nobel Prize-winning discovery, awarded to both scientists in 1993 'for the discovery of a new type of pulsar, a discovery that has opened up new possibilities for the study of gravitation'.

The Hulse-Taylor binary pulsar is in a small, high-eccentricity, short-period (eight-hour) orbit around a second neutron star. Its pulses arrive early or late according to whether the pulsar is on the near side or far side of its orbit. The timing discrepancies make it possible to map the orbit very precisely and to see how general relativity causes changes. In fact, after only two years, while Hulse was writing up his thesis in 1975, the effects of general relativity had been first seen in the pulsar timings. By 1980 it was possible for Taylor to see the full effect of the gravitational waves that the pulsar emits. The loss of energy from the system, as the gravitational waves radiate away, causes its orbit to shrink by 1.5 centimetres (0.6 inch) every orbit, so it has shrunk by 1 kilometre (0.6 mile) since it was discovered.

Further binary pulsars have been discovered and monitored since the Hulse-Taylor pulsar. In none of these cases has it been possible to measure gravitational waves themselves, only the effect of the leakage of energy from the stars' orbits due to gravitational waves. The calculations of general relativity fit amazingly well, to an accuracy much more precise than 0.1 per cent. This degree of certainty about gravitational waves inspired LIGO.

LIGO, which made the first direct detections of gravitational waves, is two instruments in Hanford, Washington, and Livingston, Louisiana, working in conjunction with a similar detector near Pisa in Italy, called Virgo. A fourth detector, the Kamioka Gravitational Wave Detector (KAGRA) in Japan, came into operation in 2020. Having multiple detectors distributed globally makes it possible to distinguish celestial events from local disturbances, like Earth tremors. Moreover, as the gravitational waves sweep through the Earth they arrive at the detectors at successive moments, which enables the scientists to estimate the direction from which the gravitational waves have come. The line of travel of the gravitational waves tracks back towards their origin.

The principle used by the instruments is to measure with a laser the distance between two freely hanging pendulum mirrors, operating as an optical interferometer. The passage of gravitational waves causes the mirrors to bob like corks on the sea, changing their separation. That alters the optical pattern in the interferometer. On 14 September 2015, LIGO saw its and the world's first detected gravitational wave event. It proved to be gravitational waves from two merging black holes. In 2017, the Nobel Prize in Physics was awarded to Americans Kip Thorne, Rainer Weiss and Barry Barish for their role in making the first detections: Thorne is an expert in the theory of gravitational waves; Weiss invented the laser technique used by the LIGO interferometers; and Barish led the project to make and use them.

The first event, GW150914 ('GW' to denote a gravitational wave event; the number is the date the event was seen, in YYMMDD format) was only two-tenths of a second in duration. The two black holes circled each other in their final orbits, quickly spiralling inwards with their orbital speed increasing. The gravitational waves made a 'chirp', with their frequency increasing from 35 revolutions per second to 250 per second, following the decrease of the orbital period as the two black holes got closer. The two black holes touched and merged into one. The resulting black hole oscillated afterwards for a few hundredths of a second.

The two black holes in this event were thirty-five and thirty times the mass of the Sun; when they merged, they produced a black hole of a mass sixty-two times the Sun's. Three solar masses had gone missing. This mass had been converted to the energy of the gravitational waves, through the $E = mc^2$ equation, and radiated away. The merger took place in a galaxy at a distance of 1.5 billion light years. Diluted by being spread out on the surface of a sphere of that radius, a minute fraction of this energy had been responsible for moving the pendulum weights in LIGO. Later gravitational wave events have been up to ten times farther away, so a large fraction of black hole mergers in practically the entire Universe lies within range of the gravitational wave detectors – with the proviso, of course, that the merger event takes place at the right moment and at the right distance so that the gravitational waves travel to arrive when the detectors are operating.

The signature of gravitational waves from the mergers was precisely what had been expected, which made analysis of the observation straightforward – the way to do it had all been well worked out, discussed openly for years and agreed in the community beforehand. This must have contributed to the speedy response of the Nobel Prize committee and the short interval between discovery and award. In that time (fifteen months) a dozen further events had been logged and all but one were mergers of black holes.

What was unexpected was that the black holes were so massive – not supermassive but more than stellar. Black holes in a binary star system (it had been understood) come from supernova explosions of the progenitor stars. Cutting a long story short, astronomers believed until then that only smaller black holes could be the result of a supernova and were expecting that they would see merging black holes with masses in the range of 5–20 solar masses, not the 30–50 solar mass range that has transpired. There must be something that astronomers do not understand – but what? The answer is a discovery yet to be made.

Nearly all the events seen (numbering well over fifty up to 2020) have been black hole mergers. A handful have been mergers of one neutron star with another, and one the merger of a neutron star with a black hole. The first neutron star merger event, GW170817, took place in 2017. The two neutron stars spiralled into each other over a period of nearly two minutes, the frequency of each revolution increasing from 24 revolutions per second to about 300. The merger seemingly produced a neutron star of up to about 3 solar masses. Considering that neutron stars can survive only if their mass is less than 2 solar masses, that meant that this new one was hypermassive and doomed. There are signs that it quickly collapsed to a black hole – the merged neutron star was just a step on the way to oblivion.

It is conceivable that just as neutron stars merge, white dwarfs might also merge to create a white dwarf star that is temporarily more than the maximum mass it can usually be. It, too, would then quickly collapse to a black hole. Such an event would look like an extra-energetic Type Ia supernova. This is a model for Type Ia supernovae that is an alternative to Iben and Whelan's model described on page 165. There is a star known as WDJ0551+4135 that appears to be the result of the merger of two white dwarfs into one. Its mass is 1.14 solar masses and it just missed out on exploding as a Type Ia supernova. By contrast, there have been Type Ia supernovae that were abnormally bright and the reason why might be that they were outbursts caused by merging white dwarfs too big to survive.

Black hole mergers themselves produce no significant light, radio wave or X-ray energy, but neutron star mergers splatter about material that picks up energy from the event and radiates in ways that optical, radio and X-ray telescopes can see. Two seconds after the gravitational wave burst from the GW170817 event, a short gamma-ray burst (two seconds in duration) was detected by the Fermi Gamma-ray Space Telescope and INTEGRAL (the INTErnational Gamma-Ray Astrophysics Laboratory) spacecraft. Gamma-ray bursts are labelled with letters and date numbers using the same convention as gravitational wave events, so GW170817 was quickly followed at the same place by a burst of gamma rays, GRB170817. The date designations are identical because the two events were almost simultaneous, within seconds of each other. Eleven hours later, a new, transient point source of light was found in a galaxy during a search of the region indicated by the gravitational wave detectors and the gamma rays.

An event like this is called 'multi-messenger' because the astronomical information is conveyed by several sorts of radiation detected by different kinds of telescope. Each radiation brings a different opportunity for an investigation, which, taken altogether, can provide the most complete picture of what happened. It is a challenge to take advantage of the opportunities offered by multi-messenger events because they take place without warning, anywhere in the sky, at any time. They are often transient, fading away in seconds or, usually at most, days, so there might be very little time for each telescope crew to get their act together. To reap the evident advantages, astronomers have tried to organize themselves into groups for multi-messenger studies, coordinating their equipment in order to observe transients. They have constructed observing networks ranging in complexity from two telescopes mounted on the same satellite and pointing in the same direction, to global networks of dozens of telescopes.

In a global network, an event detected by one telescope triggers its fellow telescopes to start their investigations. NASA's space telescope, the Neil Gehrels Swift Observatory launched in 2004, is studying gamma-ray bursts and has an autonomous system on board to detect bursts. It issues an alert within seconds that triggers other telescopes to interrupt whatever they are doing, to slew to point to the position of the burst and to bring their capabilities to bear. Of course, there are uncontrollable natural environmental circumstances that may upset plans – having a space telescope located on the wrong side of the Earth at the critical moment or having a ground-based telescope languishing under thick cloud are two common setbacks.

The phenomenon that was GW/GRB170817 lasted in its entirety for a sufficient length of time that it was successfully monitored by seventy telescopes, joining in one after another to observe with radiations from gamma rays to radio wavelengths. The gamma-ray, ultraviolet, optical, infrared and radio sources faded mostly away over the hours, days and weeks that followed the gravitational wave detection, with some unexplained weak radiations seen even two and a half years later. The event proved to have taken place in the elliptical galaxy NGC 4993, 130 million light years away. The X-ray, light and radio radiations were from an exploding, rapidly cooling cloud of material, debris ejected from the neutron star merger from which the gravitational waves and the gamma rays were emitted. The phenomenon is known as a kilonova.

The afterglow of the gamma-ray burst GRB170817 was powered by the radioactive decay of heavy nuclei generated in the event. Kilonovae produce half the chemical elements heavier than iron in the Universe, approximately 16,000 times the mass of the Earth in each event. Heavy elements include gold and platinum; it is intriguing to look at a piece of gold and platinum jewelry, like a ring, and imagine the time chain of the creation of its elements in the violence of a kilonova, their travel through interstellar space into the solar nebula and the Earth, and then the geological and human processes that brought them to your third finger. Without kilonovae to make them, these elements would be even more uncommon and valuable than they actually are.

Gamma-ray bursts

The gamma-ray burst seen from GRB170817 was a type of explosion first detected in the 1960s by US military satellites called Vela spacecraft. The 1960s was a tense time during the Cold War between the Western allies led by the USA and the Eastern bloc led by the then Soviet Union (the USSR, now the Russian bloc). Each side was developing nuclear weapons, and tests carried out on the surface of the Earth were generating dangerous nuclear fallout, which drifted around the world, troubling the innocent. The Partial Test Ban Treaty was signed in 1963, which banned tests of nuclear weapons in the atmosphere, in space and under water (but permitting underground tests). To monitor compliance, in 1967 the USA launched a constellation of Vela satellites. They were designed to detect gamma rays in short bursts, as emitted by nuclear weapons. To everybody's amazement the satellites immediately saw gamma-ray bursts at a frequency of about once per day, far too many to be due to nuclear tests. For reasons of military security, this information was not generally released until 1973. The bursts came from random directions in space, and it became obvious that they were natural celestial phenomena.

As explained above, celestial gamma-ray bursts are given a date number starting with the year, followed by the month and day. If, rarely, there is more than one burst in a given day, a subsequent letter is added – A, B and so on. Each burst lasts between a few thousandths of a second to, typically, several minutes. The bursts fall into two groups: short bursts with an average duration of one-third of a second, and long bursts with an average duration of about half a minute, although they may last for more than a couple of hours. The bursts, particularly the longer ones, are often followed by an 'afterglow' – a source of light and radio waves suddenly appears and then slowly fades away.

A short burst of 0.3 seconds' length must come from something that is less than 0.3 light seconds long. Otherwise, it would be blurred out into a longer pulse by the light travel time between the front and back of whatever it is that is flashing (0.3 light seconds is 100,000 kilometres/60,000 miles, only a few times the size of the Earth). The origin of the short bursts as two merging neutron stars was uncovered by the study of GRB170817, as described above.

The longer bursters are probably from 'naked' supernovae. These events are due to the collapse of a star that has, in the course of its earlier evolution, shed its outer layers to such a depth as to expose its core. The collapse of the core of a Type II supernova to a black hole always generates gamma rays, but usually they are absorbed by the outer body of the star. If that outer body is missing, the supernova is naked and the gamma rays are able to flow into space, because there is nothing to stop them. The gamma-ray bursts sometimes flare repeatedly in complicated and apparently random patterns. This phenomenon is thought to relate to lumps of material ejected by the supernova falling back and being eaten by the new black hole.

The connection between long bursters and core-collapse super-novae was proved when examples were found with a gamma-ray burst and a radio-wave and light afterglow all in coincidence. The afterglows were often seen to be located in galaxies, at great distances. This proved just how energetic the events were, each releasing an amount of energy comparable to a supernova. In fact, at face value, it seemed that they release hundreds of thousands of times more energy than a supernova, hence the alternative name for this phenomenon: 'hypernova'. The energy calculation was ramped down, however, when it was realized that the gamma radiation was beamed. We see gamma rays from a burst only if we on Earth are in the brighter part of the beam. If we think this bright part is typical, we consequently greatly overestimate the total energy being produced. Nevertheless, even if hypernovae are not as powerful as was once thought, they are hundreds or thousands of times more powerful than supernovae. Gamma-ray bursts are the most powerful stellar explosions known in the Universe. Only events associated with supermassive black holes, on a galactic scale, are bigger. Since we do not see gamma-ray bursts unless we are in the beam, most of them are missed and only about one burst per day is seen. For every one observed, there are five hundred more.

The light and radio-wave afterglows are caused by the stimulation of interstellar material that surrounds the progenitor star by jets that are the channels through which the energy of the hypernova is released. That interstellar material differs in its amount and distribution from location to location and thus the afterglow shows different behaviour from event to event.

The most powerful gamma-ray burst recorded is GRB080916C, which was detected on 16 September 2008 by the Fermi Gamma-ray Space Telescope. It released the energy of approximately six thousand Type Ia supernovae. It was 12 billion light years away, in a galaxy identified a few hours after the burst with a dedicated multi-messenger telescope – the Gamma-ray Burst Optical/Near-infrared Detector (GROND) attached to the 2.2-metre (90-inch) telescope at the European Southern Observatory's site at La Silla, Chile.

All these events from different types of dying stars are very powerful. Their energy derives from the release of potential energy from the mass of a star as it collapses to become a neutron star or black hole. This energy is in the region of 10^{44}–10^{46} joules, which radiates outwards into the Universe, in different forms and for a variable length of time depending on the exact circumstances of the explosions. This is a lot of energy, equivalent to that radiated by the Sun in its entire lifetime of 10 billion years or so. At just a few seconds, however, the explosion of a gamma-ray burster is brief so that all the energy is concentrated into a short, powerful burst.

If the energy of collapsing stars or merging black holes is not absorbed by something surrounding them, the burst travels for billions of light years and penetrates as a detectable amount into a significant fraction of the Universe. Mergers of stellar black holes are the most powerful such events. The flow of their gravitational wave energy is almost unimpeded by anything, and produces an event that we can see from a distance of 10 billion light years, which we could loosely describe as being on the other side of the Universe. These are the biggest bangs since the Big Bang.

9
The Birth of the Solar System

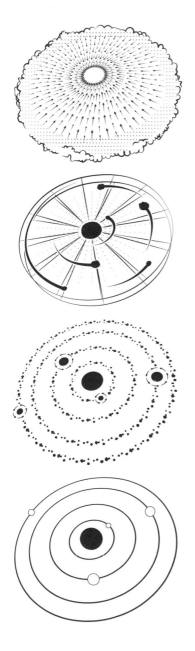

The solar system is extremely important to us as our home environment, the one that gives us our existence. The present age of space exploration and powerful telescopes has widened the compass of our astronomical knowledge and understanding of that environment. We are learning about the planets in individual detail, and can compare them, to an increasing extent, with planets orbiting other stars. This has added more and more numerous certainties to what used to be just plausible guesses about how planets and life evolved. Planetary science and astrobiology (the science of extraterrestrial life) are multidisciplinary subjects of increasing sophistication and significance.

A disc of gas and dust surrounded the Sun as it formed. Dust particles stuck together and built up the planets.

Icy dust in the solar nebula

On a cosmic scale the solar system is insignificant compared to galaxies and stars, and in cosmic history it is a late arrival, which is why it appears towards the end of this biography of the Universe. This book commenced with a Big Bang, but the story is only now reaching towards what appears to us to be its culmination: the emergence of humankind as a cosmic phenomenon. This perspective is self-centred and what appears to us to be the climax is, in a broader perspective, of limited impact and short duration (see Chapter 12).

The potential for planets to exist and to host life on their surface is present in the interstellar medium of our Galaxy, and was realized in our solar system as the Sun came into existence in the solar nebula. The solar nebula formed a disc of gas and solid grains that rotated around the nascent Sun, travelling in more or less circular orbits (see Chapter 6). The gas originated in interstellar space and was mostly hydrogen and helium, made in the Big Bang, but it also contained simple molecular compounds, made of elements like carbon, oxygen and nitrogen, themselves made in stars. Molecules of few atoms were the more common because they were simpler to make in the vacuous spaces of the Galaxy than complicated molecules. They included hydrogen molecules (two hydrogen atoms, H_2), water (two atoms of hydrogen and one of oxygen, H_2O), ammonia (three atoms of hydrogen and one of nitrogen, NH_3) and carbon mon- and di-oxide (one atom of carbon and one or two of oxygen respectively, CO and CO_2); more complex molecules like methanol (one atom each of carbon and oxygen and four of hydrogen, CH_3OH) and acetaldehyde (two atoms of carbon, one of oxygen and four of hydrogen, CH_3CHO) also occurred. The most complex molecules discovered so far in interstellar space have a dozen or more atoms, like benzene, C_6H_6, with six carbon and six hydrogen atoms arranged in a ring. Buckminsterfullerene molecules (informally known as 'bucky balls') have only one kind of atom but many of them: each has sixty carbon atoms arranged in a structure typical of a geodesic dome designed by the architect Buckminster Fuller.

We could infer that these might have been about as complex as the molecules of the solar nebula, before it became warm and dense and cooked up anything more complicated. The chemical composition of the disc was similar to that of comets in the modern solar system; indeed, the solar nebula was the material from which comets originated. Interstellar space is cold, like comets, and these molecules, and others like them, were at the outset frozen onto the grains of the solar nebula as various forms of ice, then amalgamated together as comets 4.5 billion years ago. Until a comet for some reason ventures into the warmth of the Sun, which activates further chemical reactions, it will have remained little changed since then.

At the same time as these molecules were accumulating into the solar nebula, the Sun was forming at its centre, brightening and radiating more heat. This warmed those grains that were unshadowed in the zones nearest to the Sun. There, the ice melted and vaporized – or more correctly, in space the ice 'sublimated', turning directly from solid to gas without passing through the liquid form. (Frozen carbon dioxide commonly does this even on Earth, as when dry ice is used in a gothic theatrical production to simulate fog on the stage.) Solid grains were left behind as the sublimated vapour joined the other gases in the solar nebula.

The boundary between the inner zone where most of the frozen ices are vaporized and the outer zone where they remain solid is called the snow line, by analogy with the contour line on a high mountain above which snow remains frozen all the year. Typically, the snow line in young stars like the Sun lies at a radius of two or three times the radius of the Earth's orbit (the radius of the Earth's orbit is defined as 1 astronomical unit, so 2–3 astronomical units, corresponding to about the orbit of Mars in our solar system). Inside the snow line, the solar nebula became dry, made of solid grains, and the ices vaporized. Molecules of the gases (mostly hydrogen and helium) warmed and dissipated away from the Sun. Outside the snow line, grains kept their icy cover and the gases stayed put.

The snow line differs somewhat in its location from compound to compound, depending on the temperature at which the solid form of the compound changes to vapour. 'The' snow line is taken as the boundary between water vapour and water ice. In general, however, there develops a progressive change of composition of the solar nebula as different compounds vaporize. This means that there are compositional differences in solar system bodies, depending on where they formed.

As a protostar forms within a protoplanetary nebula, the snow line is close to the star and it has not been possible with current telescopes to image the nebulae and see the boundary except in one unusual case, V883 Orionis. The image was obtained with the Atacama Large Millimeter/submillimeter Array (ALMA), which is a unique international telescope operated by the European Southern Observatory that studies the Universe from a high plateau at Chajnantor in the foothills of the Andes Mountains in Chile, near to San Pedro de Atacama.

ALMA is composed of sixty-six high-precision antennas, operating at wavelengths of 0.32–3.6 millimetres. They are connected together in an interferometer and act together as a single telescope. Antennas can be moved on huge transporters so the interferometer can be arranged in different configurations, where the maximum distance between antennas can vary from 150 metres (500 feet) to 16 kilometres (10 miles). There are two transporters called Otto and Lore that do this job, picking up and repositioning the antennas on their mounting pads to a precision of about a millimetre. This enables the interferometer to make detailed pictures of the sky by combining data from all the antennas in all of the configurations through a specialized computer called a correlator. ALMA's pictures can be sharper than the pictures from the Hubble Space Telescope.

ALMA operates by sensing millimetre and submillimetre radiation. Such signals from space are heavily absorbed by water vapour in the Earth's atmosphere, so telescopes for this kind of astronomy must be built on sites where the air is thin and dry. The ALMA site, in the Atacama Desert at an altitude of 5,000 metres (16,000 feet), is one of the driest places on Earth and because of the height, the air is thin. Workers at the site must breathe oxygen from cylinders to operate this frontier observatory, almost as if on a spacewalk, but it is worth the difficulty: ALMA has opened a new window into the Universe, through which, among other things, it can see the warm dust and the molecules in exoplanetary systems. Unfortunately, astronomers can detect only gas and dust and cannot directly see anything that has a size between a tennis ball and a large planet and so they cannot detect the stages by which planetesimals grow from pebbles and rocks (meteoroids) into asteroids and planets. Because they are the essence of planet-building, the small bits are crucial, but their precise role is a bit of a mystery.

Even with its amazingly sharp imaging properties, ALMA has not usually been able to witness the boundary marked by the snow line in ordinary protoplanetary discs because, at 2–3 astronomical units, it is so far into the protoplanetary nebula that structure cannot be distinguished from the central star. However, there has been one favourable case, as mentioned above. In 2016, the snow line in the protoplanetary disc around the star V883 Orionis lay at a distance of more than 40 astronomical units (the size of the orbit of Neptune in our solar system). The star is like the Sun, only one-third more massive, but it was temporarily four hundred times brighter than normal. It was experiencing an outburst, a sudden increase in temperature and luminosity due to large amounts of material being transferred from the disc to the star, activating the disc and then the star's surface. The disc was flash-heated by the outburst and released molecules from its icy compounds in the inner zone of the protoplanetary disc. The water that was vaporized lay within 40 astronomical units of the star. Other molecules detected as vapour included methanol (CH_3OH), acetone (CH_3COCH_3), acetaldehyde (CH_3CHO), methyl formate (CH_3OCHO) and acetonitrile (CH_3CN). These more complex organic molecules lay within their snow line of 60 astronomical units.

The formation of protoplanets

The solar nebula formed a disc of gas, ice and solid grains that rotated around the Sun. The build-up of the grains to terrestrial planets is a subject still under study but it may have happened somewhat like this. About 100,000 years after the birth of the solar system, the grains were small and the odd electrostatic charge on one could influence its trajectory among the others. When grains of opposite electrostatic charge touched, they stuck together. In another similar effect, a molecule on the surface of a grain might have an atom broken off so that an unsatisfied chemical bond waved free. If two grains with free chemical bonds touched, the bonds might link and connect the two grains, again sticking them together. In ways like these, small grains coagulated and grew, becoming centimetre-sized pebbles.

Once the grains had become pebbles several million years later, they grew faster; they sailed through the dust and gas of the solar nebula, whose material flowed around both sides of the disturbance, colliding behind the fragment and falling onto its rear. In addition, dust particles or smaller pieces hitting a fragment would break up and bounce in pieces, settling on the surface of the fragment. These processes built up the fragment, in a process called accretion. The pebbles grew to boulder-size and larger, metres to kilometres in size.

At this stage, the fragments are termed 'planetesimals'. These larger fragments moved together, attracting each other with the force of gravity to form protoplanets. Some protoplanets merged to become the planets and some remained as asteroids in the present-day solar system. One survivor is the Kuiper Belt object Arrokoth (the name means 'sky' in the Powhatan/Algonquian language and its numerical designation is 2014 MU69). Arrokoth was formerly nicknamed Ultima Thule (which signifies 'beyond the most distant lands') because of its position in the solar system beyond Pluto. The New Horizons space probe flew by Arrokoth on 1 January 2019, having passed Neptune, the most distant planet, and even Pluto. Images from the fly-by (pl. XIII) showed that Arrokoth has two connected lobes 30 kilometres (19 miles) long overall, so that from some angles it looks like a snowman. It has a smooth surface and a uniform composition, indicating it is pristine, unaltered since its origin.

The lobes of this two-part object were once separate bodies that formed close together, moving slowly relative to each other, at a walking pace of perhaps 1–2 metres (3–6.5 feet) per second. They were mutually attracted by gravity into an orbit, the one around the other. They touched, rubbed together and then gently merged. Each of the lobes is itself made up of smaller lumps that had earlier fused together. Arrokoth looks the way it does, not because it smashed violently together with collisions, but because it accumulated gently with small kisses and enveloping hugs.

The protoplanets that formed near to the Sun got bigger by accumulating material that they encountered on their orbits and by merging together. Outside the snow line, larger gas giant planets like Jupiter formed in the abundant, gaseous and dense nebular material, in the same way, more or less, that galaxies had earlier formed from hydrogen made in the Big Bang, or that stars like the Sun had formed from interstellar material in our Galaxy, by collapsing around the place where a somewhat denser clump of gas had condensed.

The gravitational force from each protoplanet became large enough to influence the surrounding nebula. Each accreted further material, feeding as it rolled along its orbit. Like a lawnmower driving forwards through long grass and filling the grass-collector, each protoplanet emptied its track through the solar nebula over a period measured in hundreds of thousands of years. The cleared tracks are a characteristic of many of the pictures of protoplanetary discs made by ALMA. They offer evidence for the existence of the exoplanets that emptied the tracks. ALMA's pictures look out into space and view other planetary systems but, perhaps more importantly, they enable us to visualize our own solar system as if we were looking back in time to its formation and the time its planets were accreting.

Eventually, some protoplanets became massive enough to count as planets, reflecting the light of their parent star. From nebula through protoplanetary nebula to protoplanets took a relatively short time as astronomical time periods go, about 10 million years.

The new protoplanets were warm from their birth process: the rain of dust, pebbles, rocks and asteroids had heated the planetary surfaces by impacting on them, and processes such as the decay of radioactive materials had released heat internal to the planets. They warmed further from the outside by collecting heat from their star. There is one system known that mimics the solar system at this stage: the exoplanetary system orbiting the star PDS 70 (pl. XII). It is a star with a mass not much different from the Sun; its two planets are jupiters, orbiting at Jupiter-like distances from their star, so if smaller rocky planets are uncovered inside the orbit of the inner planet this may have been how the solar system looked at its birth.

Within the snow line of the solar nebula, the grains were dry, solid, rocky material. Beyond the snow line the grains retained their icy coatings. Grains are made of elements that are not very abundant in the Universe compared with hydrogen and helium, so there was not much mass in the solar nebula within the snow line. So, when, within the snow line, the dry grains coagulated to form small planetesimals, they became the rocky planets. These included Mercury, Venus, Earth and Mars, but perhaps also other terrestrial planets that have since disappeared. They are all low-mass planets – the lighter and very much more abundant gases (hydrogen and its compounds, and helium) have dispersed and floated off.

Outside the snow line the icy grains amalgamated and went on to form the gas giant planets – Jupiter, Saturn, Uranus and Neptune – and again, perhaps there were others at first. These outer planetesimals formed in the densest part of the solar nebula, replete not only with solid grains but with unvaporized ices and even greater amounts of hydrogen and helium gas. The outer planetesimals dragged in the gases that lay within a feeding zone and so built up massive, gassy planets.

In a litter of animals like pigs, one, usually the biggest one at birth, is often more successful in feeding itself than the others and is called dominant. It starts bigger than all the others and gets proportionately even bigger by gobbling more food than the rest. If the solar system could be viewed in the same way, Jupiter would be the dominant planet. It successfully pulled in more than three times the mass of the second-largest planet Saturn, itself nearly six times the mass of the next largest planet. Jupiter fed on a wide zone and its gravitational pull extended even further than its reach. As a result, it stirred up the planetesimals forming in the solar nebula nearby. Many of the planetesimals collided, broke up and could not mend themselves. Thus, Jupiter inhibited the formation of a planet in the place where we now see asteroids.

The asteroid zone contains asteroids from various origins. Some are planetesimals, like Bennu. Some of them are immature, small planets – more than planetesimals and less than a full-sized planet. Ceres is one of the larger asteroids, a spherical so-called 'dwarf planet' in fact, and perhaps originated in this way. Some asteroids, like Ida and Gaspra, appear to be fractured planets because they are so angular. Collisions in the crowded asteroid belt were frequent and small pieces broke off larger asteroids and this accounts for their irregular shapes.

As Jupiter grew large by gathering infalling streamers of gas, the streamers interacted and created a disc-shaped nebula of dust and gas, centred on the planet, much like the nebula from which the solar system originated. This nebula developed into a mini-solar system, which became Jupiter's system of satellites , the four largest of which are called the Galilean satellites, after their discoverer. Some inner satellites did not survive; they fell onto the proto-Jupiter.

As the solar nebula disappeared, Jupiter's nebula became unconstrained and had room to expand. Moreover, the satellites grew by accretion, so the nebula diminished. Jupiter gathered the remaining material, and the growth of the satellites ceased after about 100,000 years. However many satellites were created at the outset, few of the original satellites are left in orbit today, the others (if there were others) having disappeared, perhaps having fallen on Jupiter or been ejected from Jupiter's gravitational control. The Galilean satellites are the largest four. Since this time of the origin of the Galilean satellites, however, Jupiter has captured numerous passing asteroids and this has brought its present number of satellites up to eighty.

The quartet are Io, Europa (pl. XIV), Ganymede and Callisto. All are 3,000–5,000 kilometres (2,000–3,000 miles) in diameter, comparable with our Moon. Distant from the Sun's warmth, and outside the snow line of the solar system, they did not grow massive enough to retain an atmosphere, but they did accumulate and retain a considerable amount of water ice. Thus, they started their lives as similar ice-rich rocky worlds. Because of the varying history of warming that they have had subsequently, they developed individual characters of their own (see Chapters 10 and 11). Jupiter has nearly eighty known satellites. Other than the Galilean satellites, most were passing asteroids that ventured too close and were captured.

Saturn may have developed its system of satellites in the same way and at the same time as Jupiter's Galilean satellites, including its largest satellite Titan and some of its mid-sized ones. Only Titan grew massive enough to retain a substantial atmosphere. Like Jupiter, Saturn went on to accumulate many further satellites: captured asteroids. It has more than eighty satellites in total, as well as innumerable tiny satellites that make up its rings.

Migration of the planets

The planetesimals and the residual gaseous parts of the solar nebula continued to interact. Each planetesimal exerted attraction on the gas both inside and outside its orbit. The interaction started by causing the planetesimals to drift inwards. If uninterrupted, this migration would have created a solar system far different from the one we actually inhabit. The Earth would have been a casualty, swept into the Sun. The gas giants may have survived but would be in orbits much closer to the Sun.

If this had happened, our solar system would have been much more like many recently discovered exoplanetary systems than it is. Our solar system consists of eight planets spread in distance from 0.6 to 40 astronomical units from the Sun (four of them massive gas giants between 5 and 40 astronomical units), orbiting with periods between 3 months and 165 years (the gas giants between 12 years and 165 years). Typical exoplanetary systems discovered so far have one or two massive planets lying at distances from their sun of 0.01–10 astronomical units, and with periods between 0.1 and 50 years. Exoplanetary systems typically have fewer gas giants much closer to their sun. In part, this is a matter of observational selection – it is easier to find systems that have massive planets in a short period orbit – but this selection effect can be accounted for and it seems nevertheless true that exoplanetary systems with a so-called 'hot jupiter' are an archetype; the first exoplanet securely identified was of this type. It was discovered in 1995 in orbit around the star 51 Pegasi by Swiss astronomers Michel Mayor and his PhD student Didier Queloz, who were awarded the Nobel Prize in Physics for the discovery in 2019.

The gas giant planets in hot jupiter systems must have formed further out in their planetary system, beyond its snow line, but have migrated inwards. They are now much hotter than they were, with their gaseous material evaporating and dissipating. Our Jupiter avoided this fate because, in some way, it reversed its course and tacked like a boat to sail against the tide, returning back towards its birthplace further out.

By contrast with Jupiter, which formed further out in the solar system than it is now and ended up closer to the Sun, the outer-most planet, Neptune, seems to have formed closer in and then ended up further away. Astronomers deduce this by trying to solve a puzzling feature of our solar system: it looks as if it has not been around long enough to form the outermost planets. The frontier of the solar system, where Neptune is now, is far from the Sun and the Sun's gravity is weak; exciting things do not often happen in this zone because things move slowly, so collisions seldom occur and are slow. Planetesimals that formed there did not grow large. Our solar system has such small planetesimals in its distant reaches: out beyond Neptune is the so-called Kuiper Belt, the home of the trans-Neptunian objects (the clue is in the name). These objects have origins that are heterogeneous, but many are like Peter Pan – they never grew up. The two-lobed asteroid Arrokoth described above is an example.

So, it appears that Neptune could not have been formed where it currently is. Unless something else happened that we haven't accounted for yet, Neptune should not exist. The solution to this conundrum seems to be that it must have been formed further in towards the Sun and moved out to its current orbit.

The same is indicated by an anomaly in the overall properties of the solar system. There is, by and large, a smooth progression in the size of planets outwards from the Sun. The ones near the Sun are the smallest, the ones in the central zone are the most massive, and then, towards the boundary of the solar system, planetary masses tail off. This progression must have had its origins in the density of the solar nebula: at a given ring in the nebula around the Sun, the more material that was there in orbit, the greater the mass of the planet that would initially form there. Of course, there would be processes during or afterwards that would reduce or increase the mass of the planet from this time but what was the starting point? What was the profile of the mass distribution in the solar nebula?

We know that any hydrogen and helium that transferred from the solar nebula into the planetesimals that turned into the rocky planets was not retained (unless chemically combined into heavier molecules, like water), but we can be fairly confident that the heavier elements in a planet, like iron and silicon, are representative of what it was born with. The idea to generate the mass profile of the original solar nebula, therefore, is to take the rocky component of each planet and add hydrogen and helium until the chemical elements as a whole match the Sun in composition, on the assumption that the composition of the Sun has not changed much from the time of its birth. If the augmented mass for each planet is spread over the area of the orbit within the solar nebula, that might give a satisfactory profile of the surface density of the solar nebula.

Although this method seems promising, it does not result in a good starting point to provide enough mass to make the planets of our solar system. It gives low surface densities for the solar nebula, with its mass too thinly spread to form the giant planets quickly. Jupiter, according to this method, would take millions of years to form, Uranus and Neptune billions of years. The indications are that the process of the formation of the planets took perhaps hundreds of thousands of years, or even less time.

In general, there has not been enough time for our solar system to develop, if the planets formed where they are now. It would have taken too long for enough material to fall together to make giant planets. Moreover, the longer the time that hydrogen and helium hang about, the more of it dissipates into space, warmed and pushed away by radiation from the Sun. The formation of the giant planets would not only slow down, but also never be a completed process.

Rather than completely giving up this approach to the theory of planetary formation, astronomers have looked at what extra feature could be introduced to make the theory work. What seems to succeed, and to fit in with other ideas about the early history of the solar system, is to suppose that the planets were formed at about halfway in from where they are now. This would compress the solar nebula into a quarter of the area, increase its density accordingly and begin the creation of planetesimals from a denser start, which would speed up the making of big planets.

There is an anomaly in this theory that is startling about Neptune. In the outer reaches of the solar nebula (it is supposed), there was a smooth drop-off of surface density with distance from the Sun, which should have resulted in a smooth decrease of planet mass. The actual progression starts off along this path: Jupiter is 320 times the mass of the Earth and Saturn 95 times, but then it all goes awry. Uranus is next in line outwards at fourteen times the mass of the Earth, but Neptune is bigger at seventeen times. If the outer planets lined up in order of decreasing mass, Neptune should be closer to the Sun than Uranus. Neptune not only moved outwards from where it was born, it is in the wrong place in the planetary line-up. This is a clue that there was some major reorganization of the planets after they had formed – Neptune was somehow shoved towards the outer fringe of the solar system as we shall see in the following chapter.

10

Chaos and Collisions in the Solar System

I

II

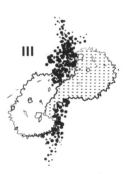

III

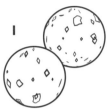

IV

Two views of the circumstances of our lives coexist in our minds, which we take out and use as the whim takes us. One is that our lives are ordered and that there is a reason for everything; the other is that everything is due to chance – in the words of Macbeth, 'a tale told by an idiot, full of sound and fury, signifying nothing'. The planets appear to move in an orderly and predictable way, but in fact, major circumstances of our existence on Earth, indeed the very architecture of the solar system, were set by chaos and chance, in unknowable motions of the planets and in accidental collisions between planets. The biography of the Earth might have been completely different.

V

VII

The Moon originated from a collision between the proto-Earth and a smaller, errant planet.

From protoplanets to the planets: the planets grow up

Early on in their history, planets become layered into zones of different composition. This is called planetary differentiation and was the result of the melting (or at least softening) and separation of different kinds of planetary materials. It started to happen in planetesimals as they grew to a size of about 1,000 kilometres (600 miles) in diameter. Denser material tended to sink down into the centre of the planetesimal and less dense material rose up towards the surface. Thus, the larger planetesimals, as well as large proto-planets and satellites, became 'differentiated' into zones of different composition with density increasing inwards. The composition of the zones depended on where the protoplanet was formed, how massive it was and the history that it had (that is, how much energy had been input into the planet and under what circumstances, to cause its material to float up or sink down).

This differentiation is one reason why there are so many kinds of meteorites. Meteorites are pieces of solid material that had been orbiting the solar system, like the small asteroids that they are, and then fell to Earth. Some of the pieces are original solar system material and have never been part of a planet of any kind. Other pieces are fragments of a broken asteroid, which had, at least partly, differentiated. The asteroid was shattered into pieces by collision with another. The mineral of which a given piece is made depends not only on the asteroid from which the meteorite came, but also on the zone within the asteroid.

In differentiated terrestrial planets, such as the Earth or Mars, or rocky satellites such as the Moon or Jupiter's satellite Io, the central region became a core (made wholly or partly of iron and nickel), surrounded by a rocky mantle and overlaid by an outer crust, and, in some cases, a gaseous atmosphere. The core of the Earth became large and layered into a solid inner core and a liquid outer core. Jupiter's icy satellites Europa and Ganymede differentiated into a similar core-mantle interior, but they developed no crust; instead, they became surrounded by layers of ice and water. Molten iron in the cores generated planet-wide magnetic fields by a dynamo effect as currents circulated in the liquid. As a result of these structures, some meteorites are made of iron and nickel and some of stony rocks, with the minerals of the stone varying considerably.

The giant planets, like Jupiter and Saturn, stratified into a deep, dense gaseous atmosphere underlain by a molecular layer of hydrogen and helium. Inside the planet where the pressure was high, the hydrogen took up metallic properties, such as are never seen in laboratory conditions because the precondition for them is a high pressure, so high that it is unattainable on Earth. These planets developed a central core, composed of heavier elements like iron, silicon and magnesium, but also of water, ammonia and methane.

The initial source of the energy that drove the evolution of rocky planetesimals into protoplanets and then into terrestrial planets was the accretion of other planetesimals, which yield up their kinetic energy in the impact to make heat. This liquified the surface into an ocean of molten magma. Residual solid material floated to the top of the magma ocean and heavier iron-rich liquid material sank. This released further energy in a runaway process that continued to separate out the material of the planet into a dense solid and liquid core, and a rocky mantle.

The liquid magma solidified on the surface as it cooled, starting within a million years after the formation of Earth. This process became complete in tens of millions of years, with the surface shielded from further meteoric bombardment by the atmosphere that it developed (except by large meteors that had the energy to break through the atmospheric barrier). Unlike the Earth, the Moon, being so much smaller, could not retain any significant atmosphere and the bombardment of the surface continued.

Further heat sources that acted later to develop the structure of the planets were the decay of radioactive elements and tidal heating, if the planet or satellite was a member of a planet/satellite pair, as with Jupiter's satellites Io and Europa. Another factor was the size of the protoplanet and the amount of insulation provided by the rocky mantle that acted to keep the heat in if it was a thick enough blanket. If the protoplanet was small, it cooled relatively faster and any melted material solidified, which halted differentiation as well as turning off any magnetic field because the inner dynamo was no longer turning.

The creation of the Moon

As the solar nebula dissipated, it left behind many planetesimals and a number of large protoplanets. They moved around the Sun in densely packed orbits, disturbed in an irregular way by interaction with passing neighbours. Collisions were inevitable, and there were two extreme kinds of collisions that had entirely different outcomes.

The first was a collision of a small planetesimal with a large protoplanet at a low relative speed – for example, as one of them overtakes the other. This resulted in the disintegration of the smaller planetesimal, some churning of the surface of the protoplanet and a hot crater – in short, in the accretion of the smaller planetesimal into the larger body. This was a way that planets grew, further such collisions overlapping onto the earlier ones. Only the very last collisions left craters as scars that are still visible.

At the other extreme, the head-on collision of two equal planetesimals resulted in the disintegration of each into many pieces. This was how many asteroids were created – some went on to be captured by planets and became moons. The broken pieces could have been of various compositions depending on which differentiated zone of the broken planet they came from. Meteorites are simply small asteroids that have fallen to Earth – the two main types of meteorites are iron or stony, depending on whether they were a piece of a frozen iron core or a rocky mantle. Indications are that one asteroid, called Psyche, the sixteenth asteroid discovered and one of the ten most massive asteroids in the asteroid belt, is made almost entirely of iron; it may be the iron core of a previous larger asteroid that lost its mantle in catastrophic collisions. An eponymous NASA space mission is due to visit Psyche in 2026.

A collision type that had an outcome important for us was neither the first nor the second above, but a kind of midway case: the glancing blow of one planetesimal on another. This alters the rotation of each and breaks off a few large pieces and many smaller fragments. This was how Earth's Moon was created. The basic scenario was put forward by American astronomers William Hartmann and Donald Davis of the Planetary Science Institute, Tucson, in Arizona at a conference in 1974, at which their idea connected with work by Harvard University researchers A. G. W. Cameron and William Ward on the dynamical properties of the Earth–Moon system – their orbits and rotation. The theory languished until 1984 when an international meeting was organized in Kona, Hawaii, about the origin of the Moon. At the start of that meeting, there was no consensus about the origin of the Moon, but by its end what became known as the giant impact hypothesis had emerged as the lead idea and has remained so ever since. The Kona conference was remarkably decisive in creating a consensus that outlined the origin of the Moon, but many variants of the main idea have been put forward and the subject cannot be said to be closed.

The main challenge in sketching out how the Moon formed is to account simultaneously for the orbit of the Moon and the rotational speed of the Earth, while also explaining their composition. The idea is that the Earth–Moon system originated from the collision of the proto-Earth, Gaia, with another protoplanet, Theia, soon after the formation of the solar system – perhaps 100 million years afterwards. Gaia was 90 per cent the size of the Earth and Theia was the size of Mars. The collision was a glancing blow and sped up the rotation of the embryonic Earth, much as stroking a hand along the equator of a geographical globe will make it spin faster. The collision left the Earth rotating much faster than now. Instead of twenty-four hours, the day was five hours long.

The collision broke up Theia and smashed the outer mantle of the Earth into small, hot pieces. The materials jumbled up. Some fell back down to Earth, some flew off into space and some entered into orbit around the Earth. A large concentration of pieces gathered together and built up the Moon, continuing the orbit.

The rocks brought back from the Moon by the Apollo astronauts in the 1970s showed that the composition of the Moon is very similar to that of the outer portions of our planet – the two are more alike than different. The mantles of the Moon and Earth are identical, distinct from the composition of meteorites from Mars and from asteroids. However, the Moon is poorer in elements that vaporize readily, such as potassium, suggesting that they may have boiled off the hot pieces from which it formed. A further large difference is that the Earth has a large iron core and the Moon does not.

The explanation put forward is that the collision of Theia with Gaia created a disc of hot material derived largely from Theia, the impactor. Both Gaia and Theia were protoplanets with a central iron core surrounded by a rocky mantle. A large fraction of the rocky mantles of the two shattered in the collision and accumulated into the Moon. As a result, the composition of the lunar rocks is essentially the same as the composition of the Earth's mantle. The two iron cores merged into one, which was acquired by the Earth, the Moon acquiring almost none.

The development of the orbits of the planets

As the solar nebula dissipated, it left behind numerous planetesimals at all stages of maturity. Some were primitive bodies that were dust and grains fused into small bodies like Arrokoth (see Chapter 9). Some were larger bodies (planets), and there may well have been more planets than the eight true planets and accompanying asteroids that there are now.

The orbits of the planets back then are not something that can be calculated precisely because of the limitations of 'chaos' in solving the equations of gravitation as applied to a number of planets. This makes the calculations about more than one planet in orbit around one sun inherently uncertain. According to Isaac Newton's analysis of two bodies in orbit one around the other (the Sun and one planet), the orbits are determined for all time: they are ellipses that repeat indefinitely. But, of course, the solar system consists of more than two bodies – in the early years of the solar system many more than two. At some level, it is impossible to ignore the pull of each planet on the others, and the orbits of planets are actually much more complex than repetitive ellipses.

In fact, the extension of Newton's theory from two bodies even to just three is intractable, let alone to hundreds or thousands. In 1887, the King of Sweden offered a prize for the solution of what came to be known as the Three-Body Problem: what are the orbits of three bodies moving under the influence of their mutual attraction by gravity? The French mathematician Henri Poincaré (1854–1912) won the prize because his analysis was the most impressive, but even he did not find the precise, mathematical solution that was being sought, and since then nor has anyone else.

What Poincaré found was that he could calculate the orbits of three bodies numerically, through laborious hand calculations on paper – but the orbits were 'so tangled that I cannot even begin to draw them', he said. Worse than this, Poincaré described in an essay on chance in *Science and Method* (1908) how when the three bodies were started from slightly different initial positions, the orbits were entirely different. 'It may happen that small differences in the initial positions may lead to enormous differences in the final phenomena. Prediction becomes impossible.'

Poincaré's work has been confirmed by modern mathematical techniques, including computer calculations of thousands of cases. In modern mathematical language, planetary orbits are 'chaotic'. If you start calculations with the planets in particular places and with particular speeds, you can calculate where they will be in, let us say, 100 million years. If you displace one of the planets by just centimetres from its assumed initial position, the planets could be, after the same time, in entirely different places. Chaotic behaviour is predictable in the short term but, in the long term, it depends so much on where you start the calculation that you cannot accurately describe the behaviour over a much longer period. Weather forecasting is chaotic in this sense. Meteorologists predict the weather, more or less accurately, a few days ahead. However, the tiny air disturbances from the flapping wings of a butterfly in Brazil, which are small changes in the starting point of the calculations, completely change the prediction. This fact of weather forecasting was discovered in 1963 by Edward Lorenz, a meteorologist at the Massachusetts Institute of Technology, who coined the term 'Butterfly Effect'; University of Maryland physicist James Yorke conceived the less frivolous-sounding term 'chaos'.

Although it is impossible to say exactly what happened to the hundreds or thousands of bodies in the solar system 3 billion or 4 billion years ago, mathematicians can describe some of what might have happened, in the same way that meteorologists can make predictions about the weather. They calculate a case with certain starting conditions, alter the start slightly and repeat the forecast, alter it again and repeat the forecast yet again, many times. They compare all the outcomes and pick up those that give plausible and frequent matches to expectations. They take the features that are common to the forecasts as indicative of reality.

The best simulation of the early solar system is known as the Nice Simulation ('Nice' is pronounced 'niece', because the work took place in 2005 at the Côte d'Azur Observatory in the French city of that name). The international group of mathematicians was led by Italian astronomer Alessandro Morbidelli. According to the simulation, what happened in the first billion years or so of the history of the solar system was like a gigantic game of interplanetary billiards or pool played by hyperactive children let loose around a billiard or pool table.

The Nice Simulation starts at a time when there were many planetesimals moving among the planets. The planets at that time included at least the four outer giant planets that we know today (Jupiter, Saturn, Uranus and Neptune) and the four inner terrestrial planets (Mercury, Venus, Earth and Mars) but, it is supposed, some more besides. Perhaps there were half a dozen of each kind. The giant planets were considerably closer to the Sun than they are now, perhaps between 5 and 30 astronomical units.

As the planetesimals moved through the solar system, they occasionally encountered one of the larger planets. Sometimes, in the encounter the planetesimal was ejected from the solar system. Perhaps this happened over time to the vast majority. These became interstellar asteroids, little worlds travelling in the darkness of space, forever far from the light of the Sun, and cold beyond its warmth. They became orphans ranging the empty spaces of the Galaxy.

The same thing may well have happened in exoplanetary systems. Occasionally one of their escaped planets looms up out of interstellar space and speeds through our solar system. A specific example of such an exoplanet was 'Oumuamua, which appeared in 2017. It was discovered by one of the US Panoramic Survey Telescope and Rapid Response System (Pan-STARRS) telescopes in Hawaii. This system consists of two telescopes, with two more planned, and each is 1.8 metres (6 feet) in size – not very large as telescopes go, but each with an unusually wide field so they can survey a large amount of sky, which they repeatedly image, saving the results. The telescopes are attached to a highly efficient data analysis system, which looks for changes from image to image, such as a new star appearing, or, in this case, an asteroid moving from one place to another.

It was impossible to see directly the shape of the asteroid that Pan-STARRS discovered, but it changed brightness as it rotated, dimmer when presenting a small area, brighter when seen face-on. It was either long and thin (dim when end-on) or disc-shaped (dim when edge-on). An imaginative idea, not quite unthinkable, was that it was cylindrical or flying-saucer-shaped because it was an interstellar rocket or spaceship. The thought that it was not a natural object was supplemented by the fact that it was moving along a path that indicated its orbit was influenced by some force additional to the Sun's gravity, such as a sail blown by light pressure from the Sun or as an engine of some sort. It is more likely, however, that 'Oumuamua was an asteroid covered in nitrogen ices that vaporized as it warmed on approaching the Sun, which caused a reaction force – a rocket engine, but a natural one. This hypothesis was supported by its colour, which was distinctly red like the dwarf planet Pluto, a characteristic of the nitrogen ices with which Pluto is coated.

'Oumuamua was caught in the act of falling at unusually high speed into the solar system from outside. Some similar visitors have already come into the solar system and masquerade as familiar asteroids. However, some of them orbit backwards, having been captured from a randomly oriented direction in space – the 2017 visitor is the first asteroid to be seen on the way in, before capture. Because the Pan-STARRS telescopes are in Hawaii, the astronomers there consulted the local community for suggestions of a suitable name. The body was named 'Oumuamua, which in Hawaiian means 'the first messenger to arrive from afar'. Unfortunately, its orbit quickly took it out of sight, probably forever. It swept into and out of the solar system like a sailing boat in windy conditions that failed in an attempt to stop at a harbour mooring.

When planetesimals were being ejected from our solar system by interaction with the planets, they kicked the planets backwards a little and so the planets gradually migrated in towards the Sun. After tens or hundreds of millions of years this altered the periods of the two innermost giant planets, Jupiter and Saturn. They resonated, with two of Jupiter's orbits taking exactly the same time as one of Saturn's: this is called a 2:1 (two to one) resonance. Working together, the two planets had a profound effect on all the rest of the planets and asteroids. Some more of them were ejected into space and the outcome for the terrestrial planets was that just four were left behind – those we know today as Mercury, Venus, Earth and Mars.

At that time, the Earth might have become interstellar – a cold, lifeless planet roving around the Galaxy like a lone coyote on the icy, vacant prairie. Fortunately for us, however, this did not happen. Our planet shifted its orbit back and forth towards and away from the Sun, ending up in the Goldilocks Zone of the solar system, where it is not too hot and not too cold, but the temperature is just the right value to make possible the oceans and the evolution of life.

There was also a profound effect on the rest of the solar system. Asteroids were swung out of their orbits. Most, perhaps more than 99 per cent, were flung into the far reaches of the solar system or interstellar space. Others jaywalked across the circular paths of the planets. Some swooped so close to a planet that they were captured, like the two asteroids that became Phobos and Deimos, the moons of Mars, and a number that likewise became some of the smaller moons of Jupiter and Saturn. Those that came too close to the planets fell on them, especially those planets inwards towards the Sun, like Mercury, and their satellites, like our Moon. They pounded their surfaces, making numerous craters – this was the event that we know as the Late Heavy Bombardment.

The Late Heavy Bombardment

In the 1960s and 1970s, the USA and the then USSR were competing in the space race, with the USA responding to President Kennedy's challenge to NASA in 1961 to land a man on the Moon by the end of the decade. The Apollo lunar landing programme was the result, with a series of manned trial missions in 1967–69, culminating in the first manned landing mission, Apollo 11, which was launched on 11 July 1969 and landed in the Oceanus Procellarum nine days later. Between then and 1972 the Apollo astronauts collected 382 kilograms (842 lbs) of lunar rocks from six landing sites and brought them back to Earth. The rocks were individually selected for their potential significance by the astronauts, picked up with tongs and scoops and put into numbered bags. They were packed under vacuum into suitcase-like aluminium containers and personally escorted back to the USA.

In parallel, the Soviet Space Agency executed a series of competing unmanned missions: the Luna programme. Luna 15 was launched to the Moon from Baikonur Cosmodrome in Kazakhstan on 13 July 1969 during the flight of Apollo 11, in an effort to upstage the American landing. It arrived in Moon-orbit on 17 July and attempted to land on 21 July but impacted into the Mare Crisium (some sources say that it crashed into a lunar mountain at an altitude of 3,000 metres/10,000 feet, but no such mountain exists, so that is not accurate). Luna 15 was intended to automatically return a sample of lunar material back to Earth, a goal that was achieved in September 1970 by Luna 16 after a year of three further mission failures.

Luna 16 settled base-down on the lunar surface on 20 September 1970. It drilled 35 millimetres (1.5 inches) into the ground and brought out 101 grams (3.5 ounces) of soil, which it put into a strong container attached to a small rocket. The rocket was fired back towards Earth, where it arrived without course corrections to parachute the container onto the grassy steppes of Kazakhstan. It was a brilliant piece of automated spaceflight. I saw the container in the museum of the Lavochkin Association, the institute in Moscow that makes space vehicles for the Russian programme for scientific space exploration. The container looked as I would have expected of something that had ended a round trip to the Moon with a fiery descent through the atmosphere and a hard thump on the ground – it was black and battered. There were two further successful Luna probes that automatically returned 225 grams (8 ounces) of lunar samples: Luna 20 in 1972 and Luna 24 in 1976. The lunar soil was from sites that had been chosen for their general geological characteristics but were otherwise selected only because they were within reach of the lander.

In 2020, China became the third country to return lunar material to Earth. In its Chang'e 5 mission, a multiple spacecraft was sent into lunar orbit and dropped a lander onto the plain of the northern Oceanus Procellarum near Mons Rümker, a raised region 70 kilometres (40 miles) in dimension, formed by volcanic activity late in the Moon's history. The site had been chosen because the geological feature is thought to have formed just 1.3 billion to 1.2 billion years ago, making its rocks much younger than the typical 3-billion- to 4-billion-year-old samples collected by Apollo astronauts. The Mons Rümker material will help scientists investigate why this area of the Moon was geologically active long after activity ended in most other lunar areas. The lander scooped up lunar soil, some of it brought up from 2 metres (6.5 feet) below ground by a drill. It packed the soil into an ascender rocket on top of the lander, which then took the 1.7 kilograms (3.7 lbs) of soil back to the orbiter and transferred it into the returner. The orbiter carried the returner to Earth, where it separated and landed by parachute on the snowy grasslands of Mongolia on 16 December 2020 with its precious, rare cargo.

Other lunar rocks, less expensively collected but completely unselected by human agency, have been found among meteorites, having fallen to Earth after being knocked off the Moon's surface by the impact of asteroids. About four hundred pieces of the Moon like this have been found, amounting to 190 kilograms (420 lbs) in total. Their origin from the Moon has been established by comparing details of their composition with the Apollo samples.

If lunar rocks are splattered into space and onto the Earth by asteroid impacts, then we would expect that there would be a two-way traffic, sometimes transporting terrestrial rocks to the Moon. In 1971, the Apollo 14 commander, astronaut Alan Shepard, spotted a football-sized rock on the lunar surface. The rock came to be known as Big Bertha, or, more prosaically, Lunar Sample 14321. Its constituents were pieces that had been mixed together and frozen by a meteor impact somewhere on the Moon to make a single rock that had been ejected to the Apollo 14 landing site. The smaller pieces were mostly lunar in origin, but one piece proved better to match Earth rocks than lunar rocks. The terrestrial piece is 4 billion years old, as old as or older than any terrestrial rock found on Earth. At some time, it fell from Earth onto the Moon, to be fused by a later meteor impact into Big Bertha.

Lunar rocks have been scrutinized in great detail, with their ages determined by looking at the rate at which long-lived radioactive elements decay. There are different types of ages associated with different elements, so they measure the time since different events in the history of the rock. These can include the time since the rock last crystallized, the time since it was last struck by an impact, the time since it was dug up and, if it has been a meteorite, the length of time that it has been exposed in space to cosmic rays.

The oldest lunar rocks are those collected from the lunar highlands, the lighter areas of the Moon. The oldest moonrock of all the ones collected is 4.52 billion years old, almost as old as the very oldest meteorites that are regarded as original material from the solar nebula. Individual rocks from the dark, flat, lunar low-lands have ages that seem to cluster between 4.0 billion and 3.85 billion years. This was when they last solidified. It appears, therefore, that the crust of the Moon was strongly heated 3.9 billion years ago, some half a billion years after the Moon first formed.

The explanation for this was put forward between 1974 and 1976 by a group of astronomers at Sheffield University in the UK, led by Grenville Turner. They suggested that the Moon had first solidified about 4.5 billion years ago. It would have been impacted then by asteroids left over from the first formation of the planets. After a period of half a billion years of relative peace, the surface of the Moon was then bombarded heavily for 200 million years starting 3.9 billion years ago, remelted and resolidified. Turner called this event the 'Lunar Cataclysm', which later became known as the Late Heavy Bombardment. The event produced about 1,700 craters on the Moon larger than 20 kilometres (12 miles) in diameter and many more that are smaller.

If the Moon suffered in this way, so did the Earth, which was as much in the firing line as the Moon, even if protected by an atmosphere. Mathematically, there would have been tens of thousands of craters larger than 20 kilometres (12 miles) produced on Earth – some would have been 1,000 kilometres (600 miles) across. They have all disappeared, eroded away by 3.9 billion years of weather. However, the composition of deep ocean sediments provides some indication that the Late Heavy Bombardment did indeed affect our planet. The composition of Greenland and Canadian sediments from 3.9 billion years ago suggests that they contain more meteoritic material than usual. This layer of sediment includes material brought to Earth in the Late Heavy Bombardment.

It might also be significant that the fossil record of life on Earth seems to have started about 3.9 billion years ago – if life had evolved before this, it may have been set back badly by the Late Heavy Bombardment and most traces of the earlier life erased. If there are surviving fossils older than 3.9 billion years, they are controversial and few in number. There has been no catastrophe as large on Earth since, so life has had a free run to evolve, although not without smaller incidents, like the Chicxulub asteroid impact (see Chapter 11).

The surface of the Moon

The main surface features of the Moon are more than 3 billion years old, the oldest the lunar highlands. They are light-coloured, rough, mountainous areas made of the mineral anorthosite, a form of feldspar derived from magma. They are covered with large craters, 50–100 kilometres (30–60 miles) in diameter, caused by the impact of meteors. These large craters often have a central peak. The impact of a meteor vaporizes and liquidizes the ground of the impact site and causes an explosion that causes the ground to surge outwards and pile rocks up at a crater wall. If the surge is reflected back strongly, it converges on the central point and raises a central peak. The walls of lunar craters are often as high as terrestrial mountain ranges.

One striking difference between terrestrial mountain ranges and lunar ones is that terrestrial mountain ranges are caused by the slow collision of tectonic plates. The collision rumples up the line of collision into folds. This raises terrestrial mountains millimetre by millimetre. It takes millions of years to form a terrestrial mountain range. Lunar mountain ranges, on the other hand, are created in a matter of minutes.

Rocks from the highlands retrieved by the Apollo astronauts are typically 4.3 billion years old, some as old as 4.5 billion years. The highlands surround a number of giant ringed craters (or so-called 'basins'), thirty of them with diameters of 300 kilometres (200 miles) or more, such as the Imbrium and Orientale basins. At the end of the Late Heavy Bombardment, there was a period starting 3.8 billion years ago and lasting for 800 million years in which basalt lava oozed up from under the surface and flooded low-lying areas. This filled the basins with black lava that has since solidified. The lava cooled into dark, flat plains covering the crater floors. Although the first observers to use telescopes to view these dark features on the Moon set aside the old myths about them (which identified them as a man in the Moon, an old woman carrying firewood on her back or rabbits, or other similar folk tales), they still mistook them for bodies of water and so replaced the old myths with a newer one, which survives in the Latin names for some lunar features, such as *mare* ('sea' – the plural is *maria*), *oceanus* ('ocean'), *sinus* ('bay'), *lacus* ('lake'), *palus* ('swamp') and *rille* ('river').

Most episodes of lava flooding ended about 3.0 billion years ago, but meteors continued to impact on the lunar surface, cratering the lava plains and highlands. Some of the younger large craters have bright rays, such as the crater Tycho, which lies towards the south pole of the Moon. The longest of Tycho's rays is 2,202 kilometres (1,368 miles) long. Material that lies on the Moon's surface weathers due to exposure to the solar wind, micrometeorite bombardment and solar cosmic rays. It becomes darker in colour and the newly formed rays gradually disappear. The rays are unweathered, white debris thrown out from under the ground at the site of the meteor impact, and similar debris exposed by boulders that have been thrown out and have fallen, disturbing surface material below their trajectory. An additional indication that the craters are young is that the rays overlie the rest of the lunar surface, running in straight lines over mountains and craters with no interruptions.

Tycho is estimated as being 108 million years old: it is the youngest major lunar crater, although there are many younger smaller ones. The Tycho meteor impact was broadly contemporary with the dinosaurs, predating the impact at Chicxulub (see page 244) by 30 million years. Lunar cratering continues at a small scale even now. Telescopes that monitor the night-time areas of the Moon see brief flashes of light that are signals that a meteor impact has happened. They occur a few times per hour and create craters perhaps a few metres in diameter. Larger craters are still occasionally being made. NASA's Lunar Reconnaissance Orbiter spacecraft has monitored the lunar surface since 2009 and has identified hundreds of new craters with diameters over 10 metres (30 feet), appearing at the rate of one every couple of days. If Earth had no atmosphere, the rate at which new craters appeared on land would be similar.

Meteoritic changes to the surface of the Moon are on the same scale as the man-made changes left from the Apollo landings and other scars left by robotic spacecraft that crashed or landed on the Moon during the space age, starting when the Soviet-era Russian spacecraft Luna 2 impacted the Moon in 1959, the first time that humankind had left its mark on another world.

The lava flows that filled the maria ('seas') were the most dramatic of the volcanic events that occurred on the Moon in the past, but there are others, smaller and more recent, which have left their traces on the lunar surface. The Hadley Rille is a deep, sinuous channel at the landing site of Apollo 15, caused by flowing lava, perhaps a lava tube whose roof has collapsed. There are small, circular, vertical pits discovered by the Lunar Reconnaissance Orbiter that appear to be above a lava tube where the roof has recently collapsed in individual places. Sosigenes A is a dish-shaped lunar depression filled with a pancake-like lava flow, thought to be 18 million years old.

All these features are small and the visible face of the Moon has changed little throughout the past 3 billion years. In *Planetary Science: A Lunar Perspective*, New Zealand-born planetary scientist Stuart Ross Taylor wrote in 1982:

> A space traveler visiting the Earth 3–4 [billion years] ago would have seen the Moon rather like it is today. The red glow of a mare basalt flood could have been visible during a particularly well-timed visit. The spectacular but nearly instantaneous production of the Imbrium or Orientale basins or of a large impact crater would require finer timing to witness.

Taylor might have added that from the Earth now we see the Moon to be almost dead, but still twitching a little.

The orbits of the planets now

At the time of the Late Heavy Bombardment, the outer planets moved outwards. Moreover, the two outer planets, Neptune and Uranus, swapped positions. Neptune became the frontier of the solar system, with Uranus moving inside its orbit. This happened because, as described above, Jupiter and Saturn came into resonance, with two orbits of the one fitting exactly into the time for one orbit of the other, and their combined influence switched the places of Uranus and Neptune in the solar system.

When orbiting in the solar nebula, Jupiter moved inwards. During the upheaval at the time of the Late Heavy Bombardment, it moved outwards again. Saturn, Uranus and Neptune also moved outwards too. However, before it settled down into its current near-circular orbit, Neptune moved in an eccentric orbit, cutting across the orbits of the other planets, jaywalking. This chaos threw the asteroids around. Some were hurled to skim near other planets, where they were captured and became moons. Some asteroids became confined into the space between the orbits of Mars and Jupiter. Others were thrown outwards towards the edge of the solar system, some of them scattered into the vast emptiness of interstellar space.

The time of the Late Heavy Bombardment was the most turbulent period in the lives of the planets. There were major catastrophes still to come to individual planets, but not a general chaos that pervaded the whole solar system for an extended time. Following this period was a time during which the solar system was tidied up. The elliptical orbits of the planets interacted so that over hundreds of millions of years they changed shape and orientation, sweeping out the space around their orbits in bands. Anything in the bands was accreted – the eight major planets cleared a zone around their orbit, feeding on everything within. After the first half a billion years of its existence the solar system changed from a nebula of gas, dust and rocks filling the entire plane of the solar system to the relatively empty area that it is now.

On the one hand we can regard the solar system as almost empty. On the other hand, the solar system is full. The total width of the swept bands within the spacing of the planets just about fills the plane of the solar system without overlap. This means that there is little or no risk that two large planets will collide, although it remains a possibility that an errant asteroid might collide with a planet, or two smaller asteroids might collide in the more crowded asteroid belt. There is no room for any further planets without a risk that the extra ones would eventually collide with a neighbour.

The situation has become rather favourable for us. Most of the small asteroids that had survived the chaos and still wandered around the solar system were swept up and captured. This removed much of the risk that Earth and the other planets would be bombarded by asteroids in the future.

Of course, the planets, even the Earth, are still at risk from stray asteroids, or asteroids that are disturbed in some infrequent close encounter. Asteroid impacts – for example, the impact near Chicxulub in Mexico that changed the global climate and at least contributed to the extinction of the dinosaurs – remain a feature of the evolution of planets. Some collisions were set in train at that turbulent time, but are yet to happen: the asteroid-moon Phobos orbits close to Mars (only about 5,800 kilometres/3,600 miles above its surface) and is approaching Mars by nearly 2 metres (6.5 feet) per century; it is likely doomed to crash onto Mars in 50 million years. However, impacts of asteroids on the planets nowadays are occasional, not a sustained, lethal bombardment.

The continued interaction of the orbits of the planets causes regular cyclic changes in the orbit of the Earth and its rotation. These changes alter the location of the warmest areas on Earth, which in turn alters the directions and the strength of the winds, and therefore, ultimately, the climate. Of course, climate is the complicated result of many processes, such as the greenhouse effect, volcanism, asteroid impact and continental drift, as well as changes in the composition of the Earth's atmosphere, such as industrially and agriculturally generated carbon dioxide. Here I set aside many of these important factors and concentrate on the astronomical causes of climate change.

The warm zone on the Earth is the range of latitudes around the Equator, but over the year the warmest zone oscillates to the north and the south, because the Earth is tilted at 23.5 degrees to its orbital plane around the Sun. The warmest zone moves from the Tropic of Cancer at 23.5 degrees North in June to the Tropic of Capricorn at 23.5 degrees South in December. This produces the annual cycle of the seasons. Additionally, the eccentricity of the Earth's orbit around the Sun causes a small but noticeable variation. The Earth does not orbit the Sun in a perfect circle, but in an ellipse, which causes the distance between the Earth and the Sun to change around the year, changing the solar flux as received at Earth and therefore influencing the temperature. People who live in the northern hemisphere may be surprised to learn that the Earth is closest to the Sun in the first week of January, the middle of winter. It is, correspondingly, furthest from the Sun in the first week of July. This alters the effect of the Earth's tilt on the seasons. In January, the summer solar radiation is stronger in the southern hemisphere than the summer solar radiation is in July in the northern hemisphere, which is why summers tend to be hotter in the southern hemisphere.

If the Earth's tilt and its eccentric orbit stayed the same forever, these annual cycles of the seasons would repeat in the same way from year to year. However, because of the interaction of the orbits of the planets, the Earth's orbit and orientation changes over time. The Earth's axis does not always point in the same direction but, in a cycle called precession, points around a cone over a period of 26,000 years. Moreover, the tilt is not constant at 23.5 degrees, its current value. The angle of the cone opens and narrows between 23 and 49 degrees over a period of 41,000 years. The eccentricity of the Earth's orbit is currently 3.4 per cent but changes between almost 0 and 7 per cent, on a timescale of 100,000 years.

The changes in intensity of sunlight arising from all these orbital cycles is complicated. They were systematically calculated by the Serbian civil engineer and geophysicist Milutin Milanković (1879–1958) in the 1920s and 1930s. For this reason, they are called Milanković (or Milankovitch) cycles. He related the ice ages to these cycles, with recent cold periods occurring approximately every 100,000 years, when all the effects combined to produce maximum cooling.

Ocean sediments and Antarctic ice cores have supported Milanković's theory. Their isotopic composition varies from layer to layer, and shows how the temperature was changing over the period when the layers were deposited. In the USA, ice cores are stored in a purpose-built facility called the US National Ice Core Laboratory in Denver, Colorado. Ice cores from drill sites in Greenland, Antarctica and high mountain glaciers in the western United States show that the Ice Age presently coming to an end began 40 million years ago. An ice age is defined by glaciologists as a period on Earth in which there are extensive ice sheets, as there are in Antarctica and Greenland now, although they are retreating. The climate grew colder during the Pliocene and Pleistocene periods, starting around 3 million years ago, with the spread of ice sheets across the northern hemisphere. Since then, glaciers have advanced and retreated every 40,000 to 100,000 years. They are on the retreat now, with Milanković's astronomical cycles being the main long-term reason, but with anthropogenic global warming also adding a sudden extra push.

Earth's first atmosphere was drawn in from the solar nebula, and was replaced by a second, from volcanoes. The present atmosphere is the third, its composition generated by life in the oceans.

11

The Earth: A World of Difference

Venus, Earth and Mars all started from about the same circumstances, looking as alike as baby triplets, but history led to the development of three individual planets, ours (we think) the most exceptional. Cosmic history made a world of difference to the Earth. How did this happen?

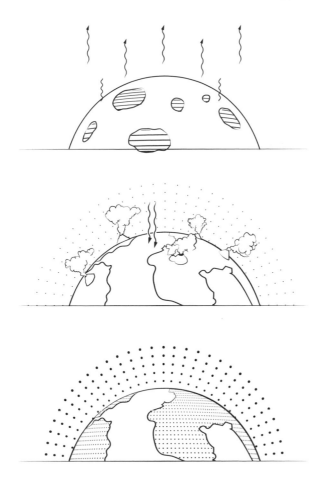

The Hadean Eon: the new-born Earth

If we were astronomers living on a planet orbiting another star, say, 100 light years away, and if we could identify exoplanetary systems, we might discover the solar system in orbit around our Sun. If, further, we could characterize the planets of the solar system to the same extent that we can characterize exoplanets at the present time, we would think that Earth, Mars and Venus are fundamentally not very different. Each is a rocky planet of much the same size, in adjacent orbits in the solar system. Yet from the much more favourable vantage point from which we can study these planets, we know how different they are. Venus has a thick atmosphere primarily of carbon dioxide over a hot, volcanic, sterile land surface, covered by sulphurous clouds. Mars, on the other hand, is a cold, almost completely dry, sandy, almost sterile desert with a thin atmosphere, also of carbon dioxide. And as we know, because we live here, Earth has a variety of solid terrains and water bodies, an atmosphere of nitrogen and oxygen and a variety of equable climates that support different forms of life everywhere, sometimes in abundance. The three planets look quite different.

How typical among planetary systems in general are the three terrestrial planets in our solar system? More than three thousand stars are known to have a planetary system. It is a challenge to learn anything about the individual planets in an exoplanetary system, but for a number of planets it is possible to estimate two of their properties: their mass (because the planet pulls its sun) and their diameter (because they block out some of their sun's light as they transit periodically across its disc). Such planets fall into two main groups: planets that are large and gaseous (like our Jupiter) and planets that are small and dense (like Earth). Deducing the structure of such Earth-like planets is certainly not as easy as adding two and two to make four; it is like adding one and one in the expectation that we might later find some further numerals to add to make four.

The currently available evidence offers up the following highly simplified picture of a typical, new-born, Earth-like planet such as our own Earth, Mars and Venus. Its dense inner part consists of an iron core, containing perhaps one-third of its mass. The core is surrounded by a rocky layer, the bulk of the solid planet, perhaps as much as two-thirds of the mass. There might be an outer layer of water that might be as much as the same mass again as the inner part. And it is all surrounded by a light gaseous atmosphere of hydrogen and helium.

During its infancy, in its first millions of years of life, the Earth developed this typical structure, modified further by impacts of asteroids (one large one in particular, when the Earth was aged 100 million years or so, which made the Moon). From this start, the Earth set off on its path to become the planet on which we live. This early period has left subtle traces in geology as the first eon of Earth's history.

In geology, the Earth's history is divided into four eons, each very roughly a billion years in length. The first, the period after which the Earth developed into its maturity, is called the Hadean Eon, named after the hellish environment of lava, volcanic eruptions and bombardments. Whether the name is apposite is questionable: Hades was the Hell of classical Greece, whose people imagined it as a dark, cold and gloomy place, not like the early Earth at all. What must have been in mind when the term was coined in relation to geology is the Hell of Christianity and Islam, which, threatened as a torment for wicked people who lived in places that are sometimes debilitatingly hot, is said to be even hotter and filled with blazing fire and sulphurous fumes.

How did the Hadean Eon come to be like this? As outlined in Chapters 9 and 10, the Earth originated 4.54 billion years ago from the solar nebula by the accretion of, at first, gas and dust and, later, by large, solid pieces. Heat from the frequent, repeated impacts of the pieces melted the impact sites. The heat was slow to leak away, and further heat was liberated inside the Earth by the decay of radioactive elements. The temperature built up and the surface of the Earth became molten magma at a temperature of thousands of degrees.

The surface of the Earth thickened with further accreted material. The outer layers acted as a blanket and trapped heat inside the surface. The interior of the Earth remained molten for millions of years. The element iron is very dense and, liquidized, it sank downwards. It took with it siderophile ('iron-loving') elements – chemical elements that at high temperatures readily dissolve in iron – including cobalt, nickel, ruthenium, rhodium, palladium, tungsten, rhenium, osmium, iridium, platinum and gold. The siderophile-iron alloys melted and percolated towards the Earth's centre.

By contrast, the lighter lithophile ('stone-loving') elements have a strong affinity for oxygen and readily formed lighter minerals that floated up towards the Earth's surface. They include: lithium, beryllium, boron, oxygen, fluorine, sodium, magnesium, aluminium, silicon, phosphorus, chlorine, potassium, calcium, scandium, titanium, vanadium, chromium, bromine, rubidium, strontium, ytterbium, zirconium, niobium, iodine, caesium, barium, lanthanum, hafnium and the rare earths (the lanthanides).

This process, in which one group of elements sank into the Earth and another group floated up to the surface, resulted in the differentiation of the Earth into a layer of a rocky mantle over a dense, mostly liquid, mostly iron core. The process is called 'the iron catastrophe', although it does not seem to be as sudden and violent as a typical catastrophe, more a slow and persistent transport of the two groups of elements in different directions. It took 10 million years to get to the stage that we see now.

The fact that the Earth has a dense core was uncovered in 1774 by the then British Astronomer Royal, Nevil Maskelyne (1732–1811), who was following up an idea for an experiment put forward by Isaac Newton. Newton imagined a pendulum, normally hanging straight downwards in the Earth's gravitational field, being positioned beside a mountain. The mountain would pull the pendulum off the vertical. The change of angle, as determined against the stars, could be measured, and that would give the pull of the mountain sideways to compare with the pull of the Earth downwards. Maskelyne chose Scottish mountain Schiehallion for his experiment because it is isolated from other mountains (which could disturb the measurements), it has steep sides so the pendulum could be close to the mountain's centre of gravity and be pulled strongly, and it is a geometrical shape whose volume, and therefore mass, could be estimated readily.

Maskelyne and his team had to observe stars to establish the vertical and to survey the mountain to determine its volume. He had to fight the weather, since cloud hung about the mountain (according to Maskelyne, its Scottish name refers to 'constant storms'). The cloud interfered with looking both up and horizontally, so his expedition took six months to complete its programme. The measurements gave the mass of the Earth, from which its average density could be derived. Modern figures give the average density of the Earth to be 5.5 grams per cubic centimetre (3.2 ounces per cubic inch) compared to the density of rock on the surface of our planet of about 3.0 grams per cubic centimetre (1.7 ounces per cubic inch). There must, therefore, be a high-density core inside the Earth in order to obtain the correct average.

In 1936, the structure of the Earth's core was uncovered by the Danish geophysicist Inge Lehmann (1888–1993) who was studying seismic waves travelling through the Earth. Some seismic waves pass from the earthquake's epicentre through the planet to seismometers located elsewhere on the surface. Their characteristics, such as speed and arrival pattern, reveal the structure of the regions through which they have travelled. Nowadays the data would be recorded digitally and analysed by computer, but Lehmann worked with written cards and pencils. She found that, underneath the rocky mantle, the Earth's core is divided into two. An inner core of iron, nickel and other siderophiles is solid, with a diameter of 2,440 kilometres (1,500 miles), a temperature of about 6,000 degrees Celsius (11,000 degrees Fahrenheit) and a density of 13 grams per cubic centimetre (7.5 ounces per cubic inch). It is surrounded by a shell of iron and nickel, a liquid outer core, with an outer diameter of 6,800 kilometres (4,200 miles) and a density of about 10 grams per cubic centimetre (6 ounces per cubic inch); this is a couple of thousand degrees cooler than the inner core.

The rocky mantle that surrounds the core is made, most abundantly, of the mineral bridgmanite, named after Percy Bridgman, a 1946 Nobel Prize-winning American physicist who studied the behaviour of substances at high pressure. Its chemical name is silicate perovskite, with a chemical formula of $(Mg,Fe)SiO_3$. It exists at the high pressures deep in the planet's interior, at depths of between 660 and 2,700 kilometres (400 and 1,700 miles). It has also been found in some meteorites, presumably originating in similar locations in other planets (or asteroids) that have been shattered by collisions, or even relating back to meteor impacts on Earth that have blasted terrestrial material into space, only for it to fall back again millions of years later.

As described in Chapter 10, soon after the formation of the mantle in the iron catastrophe, at about 4.4 billion years ago, the proto-Earth, Gaia, collided with another protoplanet, Theia, and created the Moon. This process was very energetic and melted the Earth again, at least its outer parts. Soon after that, 3.9 billion years ago, the Late Heavy Bombardment pounded the surface of the Earth and melted it for a third time.

The Moon's orbit then was much closer to Earth than it is today and there were strong tidal forces between the two bodies. The Earth locked on to a bulge that had developed in the Moon so that it was somewhat pear-shaped. The Moon kept the same hemisphere facing the Earth and still does: that is why we always see the same pattern of grey shapes on the Moon's surface that folklore calls the man in the Moon (see page 209). Dissipation of energy by the tidal forces, acting over billions of years, sucked energy from the Moon's orbital motion and the rotation of the Earth. The glancing impact by Theia on Gaia caused the proto-Earth to rotate quickly, with a day of about five hours (see page 198). Over the history of the Earth, its rotation slowed down and the Moon has retreated.

The day is still lengthening. Broadcast time (for example, the time signal heard as the 'pips' on BBC radio) is derived from

atomic clocks that keep an accurately consistent time, so it gradually diverges from astronomical time derived from the Earth's
rotation. This is the reason why, by international agreement, extra
leap seconds are sometimes inserted into the usual time sequence
of broadcast time, when needed, at the average rate of an extra
second every two years. This process resets broadcast time better
into synchronism with the rotation of the Earth.

Earth's atmosphere and the creation of the seas

The first atmosphere of the Earth was made of the lightest elements
brought in from the solar nebula, namely hydrogen and helium and
other interstellar gases. These would have included, no doubt, also
so-called noble or inert gases heavier than helium such as neon.
Helium and neon are the second and fifth most abundant elements
in the Universe, but they do not combine chemically with anything,
so they never get anchored to solids or liquids. They are always
gases that are very lightweight and readily escape into space – there
is now no more primordial helium left in Earth's atmosphere at
all, and only slight traces of neon.

In any case, the solar wind, the Earth's heat, volcanic outgassing and possibly a collision between the proto-Earth and a giant
impactor (a Pluto-sized planetesimal or larger – possibly the one
that created the Moon) transformed this first atmosphere through
an abundance of heat and vigorous chemistry. The first atmosphere
was replaced with the second, an atmosphere of hydrogen together
with carbon and hydrogen compounds, like carbon dioxide and
steam, emitted by volcanoes and also volcanic processes driven by
the heat of the ocean of magma, which still covered the Earth at
that time, or the collisional energy of the giant impactor. Smaller
planetesimals, comets, meteorites and asteroids of all sizes impacted
into the magma ocean, melting and vaporizing any icy minerals like
hydrogen cyanide, which they brought in from the solar nebula,
all contributing further to a rich gaseous mixture. Methane and
ammonia were also constituents of the atmosphere, but these are
unstable when exposed to ultraviolet light, so this atmosphere was

thus smoggy, with particles generated from organic molecules produced by the action of the Sun's ultraviolet light on the gases. In the solar system, the only other place with a similar atmosphere now is Saturn's satellite Titan, although it is far from the Sun and much colder than the Earth's atmosphere at that early time. There is no free oxygen in the atmosphere of Titan, nor was there in the atmosphere of the Earth then.

The composition of this early terrestrial atmosphere created a strong greenhouse effect, which compensated for the reduced luminosity and warmth of the Sun at that time: even though the Earth's temperature dropped as the heat of its formation radiated away, its second atmosphere prevented the surface of the Earth from freezing.

This all occurred in the first eon in Earth's history, the Hadean, and lasted from the birth of the Earth 4.5 billion years ago until approximately 4 billion years ago. The violent events that occurred during this eon scrambled the surface rocks of the Earth, burying most, so that its first half-billion years of geological history are very difficult to decipher, and presumably always will be. No, or very few, fossils are known from this time. Although absence of evidence is not definitive evidence of absence, and I have described reasons why the evidence might have been destroyed, concealed or muddled, the inference is that life had probably not yet emerged on Earth, although in the future exceptions might be discovered.

In the last years of the Hadean Eon, the Earth recovered from the impact that created the Moon and from the Late Heavy Bombardment, and prepared for the evolution of life. The surface of the Earth solidified. Water vapour was liberated from the planet's interior, produced by outgassing from rocky materials caused by strong volcanic and meteoric activity. The temperature cooled enough to condense the water, creating warm lakes, seas and oceans. Ice from comets that fell on Earth from the rest of the solar system melted and brought further water, topping up the oceans. Oceans covered a substantial fraction – perhaps all – of the Earth's surface, as now; by some sort of coincidence the volume of water on Earth is about the same as the total volume of the depressions in the Earth's crust. However, the highest mountainous areas rose above the ocean's surface. By contrast, Jupiter's satellite Europa is covered everywhere with an iced-over ocean to an average depth of several kilometres, a colder vision of Earth at the end of the Hadean Eon.

The weather in the Hadean Eon would have included rain as well as thick clouds, and strong winds were driven by large temperature differences. The eroding effects of the atmosphere on rock formations through the effects of rain and wind had begun in the Hadean almost from the birth of the crust of the Earth. Small grains of rocks that had broken away from the main surface features were blown by the wind to accumulate into more sheltered places and were flushed by streams into lakes and seas to build up new rock strata. Volcanic eruptions pumped ash and pulverized rock into the atmosphere to cause new deposits in drifts. Lava flows from vents filled depressions to overflowing. These processes continue even now, erasing the older features of the landscape and covering them up, ceaselessly rewriting the geological record of the Earth's history.

For the planet Venus, a similar resurfacing was very dramatic. NASA's Magellan spacecraft imaged Venus during a four-year mission (1990–94) to map its surface using radar. It found that Venus is currently entirely covered with a volcanic landscape. Lava has flowed from numerous volcanic craters but the lava plains are punctured by very few meteor craters. Calibrating the surface density of meteor craters on Venus against other worlds like the Moon to estimate how much time must have passed to produce that many, planetologists estimate that the entire planet was resurfaced between 1 billion and 500 million years ago. Volcanic activity on Venus in that period was stupendous but the cause of the outbreak is unknown.

Although the Earth was probably virtually lifeless in the Hadean Eon, life may have been possible in or near hydrothermal vents under the seas, deriving energy not from sunlight but from water heated by volcanic activity. Jupiter's moon Europa might be similar.

Like its neighbour Io, which has many active volcanoes, Europa (pl. XIV) is still being heated by tidal forces from Jupiter. As Europa moves in its eccentric orbit closer to and further from its parent planet, it moves between a strong and a weak gravitational force. Its material structure is 'worked' – that is, it is alternatively stretched and compressed. The interior of the satellite gets hot, not so much as Io, but to a certain extent. Heat builds up and melts Europa's surface ice from below: underwater volcanic activity is probably another result. Life on Earth may have first emerged in a deep ocean trench with an environment like this. It looks as if Europa would be worth exploring as a potential home for extraterrestrial life, but the hard part about finding life on Europa will be the journey through a kilometre of ice in order to dip into its ocean, then to get out with samples, having dived some further kilometres to the ocean floor. However that works out in the future and illuminates the origin of life elsewhere, the end of the Hadean and the beginning of the following calmer eon, known as the Archaean, seem to have been when life got under way in the seas here on Earth.

Earth's magnetic field and the magnetosphere: our shield against the capricious Sun

For centuries, European and Chinese sailors knew that lodestones (pieces of the mineral magnetite that constitute a magnetic compass) indicated the general direction of the north. If freely suspended – for example, floating on cork on the surface of water – a lodestone makes it possible to navigate a ship if the shore or the stars are not visible. The magnetic properties of the lodestone can be transferred to an iron needle for greater clarity of direction.

In 1576, an English ship's instrument maker, Robert Norman, noticed that a magnetized needle not only turned to point north, but also tended to dip down below the horizontal. The angle of dip was about 70 degrees in London. In 1600, the English physicist William Gilbert realized that this was because the needle was following lines of magnetic force that slope down into the Earth, and which extend up and out into space.

Between 1698 and 1700 the English astronomer Edmond Halley (1656–1742) combined his own magnetic survey of the Atlantic Ocean with other people's measurements to produce the first map of the world showing the direction in which a magnet pointed at any given location. His map showed that the magnetic field of the Earth is approximately a dipole, similar to that of a bar magnet, with its magnetic poles near to the geographic poles. A better approximation is possible by tilting the bar magnet by about 10 degrees from Earth's rotation axis. The simple, theoretical dipole magnetic field centred on the Earth's centre and most closely approximating the actual magnetic field has north and south poles that lie at places on the Earth's surface called the geomagnetic poles. The North Geomagnetic Pole is in Ellesmere Island, in northern Canada near Greenland. The South Geomagnetic Pole is in Antarctica, not far from the Russian Vostok Research Station.

The actual magnetic field of the Earth is more complicated than a simple dipole and it is not centred exactly on the Earth's centre. At what are defined as the magnetic poles, the actual, not the theoretical, magnetic field points directly into or out of the surface – the dip angle is 90 degrees. The North Magnetic Pole is in the Arctic region, the South Magnetic Pole in Antarctica.

The magnetic field is generated and sustained in the liquid outer core of the Earth by a dynamo process fed by motions of the liquid. The motions are driven by heat escaping from the inner core below, which causes convection of the molten iron, and by rotation, which drags the molten iron into a spin. There is also a swirling effect that is due to the scraping of the liquid outer core dragged against the solid inner core. The strength of the field at the surface of the Earth varies from region to region by a factor of two, with an average of 0.5 Gauss. There is a particularly large weak spot called the South Atlantic Anomaly, stretching from northern Chile to Africa's southern Cape. There must be corresponding variations in the temperature and density of other physical features in the flow of molten iron under the Earth's surface, for example a bump on the underside of the mantle.

The long-term history of the Earth's magnetic field is written in old rocks. They preserve residual traces of the Earth's magnetic field at the time that they solidified, provided they have been minimally altered since then. Rock specimens of different ages found at the same location are magnetized in different directions. They show that the direction of the axis of the Earth's magnetic dipole does not stay the same all the time, as English astronomer Henry Gellibrand discovered in 1635. This means that, although the geomagnetic poles do not move much, the geographic locations of the magnetic poles have considerably varied over the centuries, moving erratically by about 15 kilometres (9 miles) per year on average.

For some reason, the North Magnetic Pole has been moving much more quickly than average since 1990, at about 50 kilometres (30 miles) per year. It has moved northwards from Hudson's Bay in Canada and now lies in the Arctic Ocean. In 2017, it crossed the International Date Line close to the geographic North Pole and has now tracked from the sea north of Alaska into the sea north of Siberia. Likewise, the South Magnetic Pole has moved off the continent of Antarctica and is in the Ross Sea, just offshore in the direction towards Australia.

The magnetic field of the Earth extends to thousands of kilometres above the Earth. It is a structure called the magnetosphere, which acts like a bottle, surrounding us and shielding us against high-energy charged particles in the solar wind. The Earth's magnetosphere deflects most of these solar particles. If this was not the case, the cumulative effect of radiation might have prevented life on Earth. Even so, the Earth is frequently hit by energetic solar flares, particles from which leak into the magnetosphere causing geomagnetic storms, the aurorae and other electrical and magnetic effects (see Chapter 7).

As discovered by US geophysicist James Van Allen (1914–2006) in 1958 from the first American space probes, Explorer 1 and Explorer 3, charged particles within the magnetosphere form a doughnut-shaped region that encircles the Earth. A second, outer ring was identified later in 1958. These regions of radiation were named the Van Allen Belt in honour of their discoverer and were the first major scientific discovery to be made as a result of space exploration. Because human beings and delicate electronic equipment are susceptible to dense radiation, the Van Allen Belts, and similar regions, are factors influencing the planned trajectories of spacecraft, particularly manned spacecraft, which have to traverse them to get out to the Moon and beyond.

The early magnetic field shielded the Earth's atmosphere, preventing both air and surface water from being removed from the planet, even at a time when the solar wind was more intense. It is thought that the Earth's magnetic field is currently generated in the interaction between the inner solid core of the Earth and its liquid outer core. The inner core appeared about 565 million years ago.

The present configuration of the magnetic field of the Earth is thus a phenomenon of our planet's maturity. It has not, however, settled into inactive old age. Not only does the magnetic field move in its orientation, with the poles tracking underneath the Earth's surface as described above, but also from time to time, it undergoes a more drastic change: it changes polarity (the North Magnetic Pole changes to the position of the South Magnetic Pole and vice versa). It changes by somersaulting quickly, turning upside down.

The timing of the changes does not show a clear pattern. It seems that there are geomagnetic epochs of about half a million years during which the polarity is predominantly as now, and which alternate with epochs with the reverse polarity, switching back and forth. But there are also briefer periods of perhaps tens of thousands of years when the polarity makes a brief excursion from the predominant direction at that epoch to the reverse, and back again relatively quickly. Nearly two hundred magnetic field reversals have been identified as occurring during the past 80 million years, some more firmly established than others.

The most recent switch of major geomagnetic epochs was about 770,000 years ago (the Brunhes–Matuyama reversal, named after French and Japanese geophysicists of the early twentieth century). The best established, most recent, minor excursion took place

42,000 years ago and lasted 800 years. It is known as the Laschamp event, after the French location in the Massif Central where rocks were found with the magnetic traces that revealed its existence.

The history of the reversals shows as stripes of alternate magnetic polarity running parallel to the Mid-Atlantic Ridge in rocks of the Atlantic seabed, laid down as the ocean floor spreads out from the ridge towards both the east and west shores of the ocean. The rocks 'remember' the terrestrial magnetic field at the time they solidified and then get pushed outwards away from the ridge by subsequent lava emissions.

The time that it takes to complete a reversal is controversial, with some estimates given at several thousands of years, while others suggest just a few human lifetimes – perhaps it varies from reversal to reversal. For some period of time during the reversal, the Earth's magnetic field is much weaker, perhaps just a few per cent of its normal value. At that time, the atmosphere and the surface of the Earth are more exposed to the solar wind of cosmic rays and to high-energy radiation associated with increased auroral activity; it is not known what happens temporarily to the natural environment at that time, but whatever happens it must be survivable because reversals have happened often in the past and life persists here.

The usual time between major reversals is about 500,000 years, but the last major occurrence was nearly 800,000 years ago, so the next reversal is somewhat overdue. Are we then all doomed? No, but there may well be selective consequences. An Australian study in 2021 pointed out that the Laschamp event happened at the same time that Neanderthals and Australian marsupial megafauna became extinct. People of the species *Homo sapiens* took to living in caves about that time, perhaps motivated to shelter because of the risk of severe sunburn. Their living arrangements favoured the development of cave art, which germinated at the same time. When people re-emerged into an outdoor life, they may have been relieved to find that encounters with diprotodon (wombats the size of a rhinoceros) were no longer a danger, or perhaps they were disappointed to have lost a prey that they had formerly hunted for food.

Whatever the truth of such imaginative speculations, the oldest traces of the Earth's magnetic field survive in rocks from the northern part of South Africa that are 3.45 billion years old and possibly even in rocks from Australia at 4.0 billion years old. Our records of it thus start from the time of the catastrophic collision that created the Moon, after the circulation of the liquid iron core had recovered some semblance of orderly behaviour. The oldest magnetic rocks show that the Earth's magnetic field strength as the Hadean ended was usually comparable to its strength now, so the terrestrial magnetic field is a long-lasting phenomenon, if highly variable.

Mars: Earth past and yet to come

Mars (pl. xv) started by following a similar evolutionary path to the Earth, but unlike the Earth, it lost the water that it had at first – Mars is now dry. Most of the surface of Mars is rocky desert and large parts are covered with fields of sand dunes. Other landscapes on Mars, imaged by cameras parachuted onto the ground from space probes, show plains of rocks and dust – the surface has been broken by meteor impacts, with impact fragments scattered everywhere. There is some wind erosion but the rocks retain their hostile-looking, angular, fractured shapes for a long time.

Mars has an orbit not much different from Earth's and, potentially, it has a similar climate, but Mars is, in fact, dramatically different from our planet because its water and air have dispersed into space. Its atmosphere is now very thin: atmospheric pressure on Mars is about 1 per cent of the atmospheric pressure on Earth. By contrast to Earth, whose atmosphere is now three-quarters nitrogen, the air of Mars is more than 95 per cent carbon dioxide, even though the two planets began with similar atmospheres. The surface conditions on both planets were, for the first billion years or so, very similar. The change that took place on Mars about 4 billion years ago has preserved more of the landscape of Mars from that time, whereas the Hadean landscape of the Earth has been completely eroded by weather and tectonic activity.

During the first geological era of Mars, its landscape was flooded (the period is called the Noachian era, a reference to Noah and the Biblical Flood). How do we know what Mars was like back then? Some Martian deserts are littered with rocks that have been rounded by flowing water, tumbled like the boulders in the bed of fast-flowing streams on Earth, their sharp edges softened. Furthermore, some cliffs on Mars show strata made of minerals that are made only in standing water. There are also geological formations that were shaped by water, craters and rift valleys that show flat floors of water-deposited sediment, and river valley systems that are now dry river beds. NASA's Mars Perseverance rover landed in 2021 at the Jezero crater, which is estimated at 3.8 billion years old, and showed two breaches on opposite sides of the crater's circular wall where a river had flowed into and out of it: the floor of the crater reveals where the river deposited sediment. (This is the main reason for the choice of Jezero as the landing place for Perseverance: it is a good place to look for signs of past life on Mars.) Some Martian river systems have characteristics that indicate that the valleys were carved by the flow of ground water beneath a protective cover of ice rather than by runoff of rain. In other words, the rivers flowed under glaciers.

The water that was present on Mars was there in prodigious quantities. In the area around the Ares Vallis in Chryse Planitia, surging floods scoured the surface and formed streamlined islands by parting to go either side of the walls of meteor craters – 10 million cubic kilometres (2.4 million cubic miles) of water flowed past, making scarps around the craters, which are 400–600 metres (1,300–2,000 feet) high. Similar formations have been found on Earth, in Washington state in the USA, caused by a collapsing ice dam that released the water from a lake, and on the seabed in the Straits of Dover, which was scoured by a mega-flood from the North Sea when the land bridge from Dover in England to Calais in France collapsed.

Mars still has polar caps of water ice under carbon dioxide frost, which is deposited in seasonal annual layers. The sand dunes that surround the polar regions are stabilized by frozen ice during the winter, but in the spring, the ice loosens its grip and land slips down the sides of the frozen dunes, scraping the surface. The landslips cause billowing clouds of red dust and screes of black soil on the plain surrounding the steep edge of the Martian ice cap.

Is there still water on Mars? This a question of active research and exploration. There are some indications that spring water flows from some underground caverns and that there is a larger reservoir of water under the polar caps. If the evidence holds up to future investigation, perhaps during the first missions by astronauts to Mars, this water holds the enticing scientific prospect that any life that evolved on Mars in the Noachian era might still survive.

What altered on Mars to terminate the Noachian era? The planet's climate changed when the atmosphere was lost because of a Martian global catastrophe occurring 4 billion years ago. One reason for this loss of atmosphere was the Late Heavy Bombardment (see page 204), with meteor impacts heating the Martian air. Because Mars is so much smaller than Earth, its gravity is correspondingly weaker, so air molecules, warmed by the Sun, can more readily escape. A second reason is that Mars lost its magnetic field. The interior of Mars was initially much the same temperature as the Earth's, heated during the planet's formation by meteor impacts and by the decay of short-lived radioactive elements, but it cooled faster – its core froze only a billion years after Mars was formed and its magnetic field decayed away. It seems that Mars lost its magnetic field at about the same time that the Earth's magnetic field was re-establishing itself after the creation of the Moon.

The reason that Mars lost its heat faster than the Earth did is the same as the reason that you can eat small, baked potatoes before big ones. Large, baked potatoes taken from the oven retain their mouth-burning temperature longer than small ones: heat is lost more rapidly by a small baked potato in proportion to the internal store of heat. It is also for the same reason that, supposedly, larger birds and animals, in species like penguins and moose, are better able to populate cold environments near the poles: they retain body heat better. This latter observation is known as Bergmann's rule, after the German biologist Carl Bergmann, who formulated it in 1847 (some recent scientific discussions do not really support it, suggesting that not only physics but also biological and evolutionary factors are also at play). Whatever the status of Bergmann's rule, the small size of Mars caused it to cool much faster than the Earth and its iron core froze.

Earth has a solid inner core with a radius of 1,200 kilometres (750 miles) and a liquid outer core extending to a radius of 3,400 kilometres (2,100 miles), as large as Mars itself. Evidence about its structure comes from using earthquakes as probes. Seismic instrumentation has been deployed on Mars since 1975 when NASA's Viking landers landed on Mars; most recently, a seismometer deployed by the InSight rover has been operating since the end of 2018. However, marsquakes are both weak and rare: only a few hundred are recorded per year, measuring 2 to 4 on the Richter scale. By contrast, on Earth there are a million earthquakes per year of magnitude 2 and up, with the largest each year typically being magnitude 7, with a magnitude 8 earthquake every decade or so. The marsquakes are so weak they do not penetrate into the core of Mars and have been measured only from the single location where the InSight rover landed. As a result, there has been no study of the core of Mars in the manner of the one that Inge Lehmann made of the Earth, so there is much less direct evidence about the core of Mars than about the Earth's. Figures for its size and structure come from theoretical calculations and these suggest that its core region is 1,800 kilometres (1,100 miles) in radius, one-quarter the size of Earth's.

The mostly iron cores of the two planets are very similar in composition, but all the core of Mars is now solid. There are residual traces of a magnetic field in the rocks on the surface of Mars, so its core was once liquid with circulatory motions to generate a magnetic field. When the core solidified, the magnetic field of Mars collapsed, allowing the solar wind to scour the atmosphere and weaken it; air and water escaped. Life, it appears, ceased to develop on Mars, if indeed it had started. Fortunately for us, the Earth's magnetosphere remains strong and permitted life to develop on Earth from about 3.5 billion years ago, perhaps a bit before, and flourish since then.

The Archaean Eon: life emerges on Earth

Back on Earth approximately 4 billion years ago, perhaps a little earlier, the Hadean Eon ended. The Moon had formed, the bombardment of meteors had ceased, the surface magma ocean had cooled and spasm due to the iron catastrophe had died away. The Archaean Eon began and lasted until approximately 2.4 billion years ago. Most of what survive of the oldest rocks on Earth date from the beginning of this eon.

Early in the history of the Earth, the mantle (the layer wrapped around the core) was hotter than now and therefore more plastic. Floating on top of the mantle is the lithosphere of solid crust, the lower areas submerged under the oceans. The lithosphere was crazed into about eight plates. Light rock erupted in the middle of the plates to make thick piles that were buoyant. These were the first continents. What survives of these pieces are the strongest parts and hence they are called cratons (from the Greek *kraton*, meaning 'strength'). The oldest rocks are greenstones found in cratons in present-day Canada and Greenland, dating from about 4.1 billion years ago.

Greenstones have similarities to present-day sediments that are found in oceanic trenches. The oldest rocks show evidence of having survived high temperatures. They also include grains from sedimentary rocks that have been rounded during transport by flowing water. The conditions under which they formed hint at the environment at that time.

Convection of the material of the mantle at the centres of the plates drove tectonic material outwards but it was not until the tectonic plates had cooled enough that the outflowing edges were dense enough that they were able to sink below the edges of the adjacent plate ('subduction'). This became possible about 3.0 billion years ago and created the system of plate tectonics, in which plates are mobile and push together, setting off earthquakes and generating weak spots that become volcanoes. The motion shifts the continents around, pushed by motions of the mantle, such as the upthrust of plumes of magma at mid-ocean ridges that force apart the American and the European and African continents sitting on the plates to the west and east of the Atlantic Ocean.

Samples of rocks deposited in the Archaean Eon survive as cratons, but so much has happened geologically in the past billions of years that the disposition of these rocks as land over the Earth is unclear. The oldest evidence for the disposition of the first continents or supercontinents is for the existence of configurations called Ur (a German prefix meaning 'original' or 'primal'; fragments of this continent survive in and around India) and Vaalbara, a compounded word based on the Kaapvaal Craton (in South Africa) and the Pilbara Craton (in Western Australia), the only surviving fragments of crustal rock this old (3.6 billion to 2.5 billion years ago). These continents, formed about 3.1 billion years ago, were small, perhaps the size of Australia today.

Over the Earth's geological history, the continents have fractured and re-joined, so that the present-day continents are reassembled broken pieces. It is possible to relate areas on one continent to areas on another from the continuity of rock types and fossil species across the edges. This offers geologists the challenge to track the way that continents have formed and re-formed from these pieces, like an ever-changing jigsaw. The early history is too complex and uncertain for me to attempt to recount it briefly here.

Amid all this continuing geological turmoil, the earliest life winked into existence early in the Archaean Eon. It was in the simplest possible form of creatures (organisms with one cell without a nucleus), grouped together as prokaryotes, within which there are two sub-groups: bacteria and, sitting by their side, a group named after the eon, Archaea. Archaea survive today – in abundance among the biota in the guts of all animals, for example, but also in extreme environments otherwise inimical to life such as hot springs or salty lagoons. They are hardy. If only Archaea could organize, we might think, they could compete for dominance of the planet, over all its surface and throughout geological time. In fact, they have organized, having come together in multicellular structures such as our species and others that compete for dominance today.

The earliest clear pieces of evidence of life on Earth are fossil stromatolites – columns of bacterial mats, fossils of which from 3.5 million years ago have been found in sandstone from an ancient ocean in Western Australia near Pilbara. Less secure evidence has been offered for earlier dates for biologically generated minerals such as graphite in 3.7-billion-year-old rocks from southwestern Greenland and 4.1-billion-year-old rocks in Western Australia.

Stromatolites are formed when microbes living in the sea known as cyanobacteria bind into organic-rich sediments or precipitate minerals in alternating layers. Because the microbes are photosynthetic, deriving energy from sunlight, they progressively move up through the deposited layers towards the light, forming new layers on top of the older ones. The older layers harden into rock that grows into a stratified column or even more complex structures. When fossil stromatolites are sectioned, they show a honeycomb-like structure as the plane of dissection cuts through draped layers of mats. Cyanobacteria survive to the present day and are commonly known as blue-green algae, sometimes appearing in abundance as an algal bloom during warm summers, even in British seas. Stromatolites persist even now.

Cyanobacteria are bacteria that sit alongside Archaea as pro-karyotes, monocellular structures that were the first forms of life. Somehow, in a way that has not been identified, life emerged from non-living chemicals. It is presumed that simple organic compounds, such as amino acids, were built up from even simpler molecules and there is experimental evidence for this. In 1953, American biochemists Stanley Miller (1930–2007) and Harold Urey (1893–1981) showed that such molecules could be made from a mixture of water, methane, ammonia and hydrogen by using electric sparks to simulate lightning. The experiment has been replicated dozens of times with simulated atmospheres that more closely resemble what is thought now about the early Earth and with alternative sources of energy. They all resulted in organic molecules. Similar molecules were made in the solar nebula and could have been brought to Earth by the fall of comets, asteroids and meteors, like the Murchison meteorite (see Chapter 6). These organic molecules grouped into structures that took on properties of metabolism and self-replication and created the opportunity for growth and evolution.

Although Miller showed that the desired end result actually came about, he did not successfully identify in detailed chemical terms the way that it occurred. Soon afterwards, in 1959–62, the Catalan biochemist Joan Oró (1923–2004), working in the USA, identified the chemical reactions by which one of the gases thought to be abundant in Earth's early atmosphere – hydrogen cyanide – could develop towards more complicated molecules like amino acids and nucleic acids, as Miller had found. Oró's breakthrough has been developed into a plausible, comprehensive scheme created by molecular biologist John Sutherland and his colleagues at the MRC Laboratory of Molecular Biology in Cambridge, England. The Cyanosulfidic Photoredox Network sets out in detail how biochemistry could have developed, starting from hydrogen cyanide and water delivered to the Earth from comets and energized by ultraviolet light from the Sun. The network simultaneously generates sugars, lipids, amino acids and ribo-nucleotides, the four basic chemical molecules that are required for life to operate. These biochemicals almost literally emerged 'out of the blue'.

The Great Oxygenation Event

It is still true that there is a chasm between making biochemicals and linking them together into living organisms. However this chasm was bridged, life certainly started: the first organisms, like cyanobacteria, emerged to colonize Earth. They photosynthesize, using sunlight to activate a chemical reaction that provides energy and body mass for the organism to live. The reaction releases free

oxygen, eventually in such a quantity that this side product gradually changed the Earth's atmosphere.

At first, all the oxygen was soaked up as soon as it was liberated.
Methane, ammonia and similar chemical gases in the atmosphere
had a great affinity for oxygen and processed it into other chemicals. Likewise, the oceans contained chemically active substances
such as iron liberated by weather and transported by rivers from
continental rocks into the sea. The iron was oxidized and deposited
on the seabed. It appears nowadays as flamboyant red layers alternating with quartz, chert or carbonate minerals in so-called banded
iron rock formations. The layers are typically metres to hundreds
of metres thick and perhaps hundreds of kilometres in extent. The
layers are the source of iron ore mined commercially to produce
the metal by reversing what nature has done over millions of years
by removing the oxygen, combining it with coal (carbon). Thus,
cyanobacteria floating in oceans up to 2 billion years ago are the
basis on which modern-day heavy industry is founded.

By about 2.4 billion years ago, more oxygen was being produced
than was being soaked up in the surface of the Earth, marking
the Great Oxygenation Event – for the first time the atmosphere
contained free oxygen. It was less an event than an era in which
the oxygen level gradually increased – for example, banded iron
formations continued to be made, but only for about half a billion
years in the deep oceans into which atmospheric oxygen did not
at first penetrate.

From this time, Earth was proclaiming to any astronomers in
the Universe that it had become an abode of life, using the oxygen
atmosphere as a signal – if there is life on an exoplanet, this is the
principal way that astronomers on Earth hope first to detect it.

The Proterozoic Eon: life becomes a planetary force

The oxygen in the Earth's atmosphere became more and more abundant, as life took over the surface of the Earth and became a planetary
force. Today, the Earth's atmosphere is still nitrogen-rich and retains
some argon and carbon dioxide – nitrogen accounts for 78 per cent

of the atmosphere and argon 0.9 per cent; the change generated by the Great Oxygenation Event was that carbon dioxide has dropped to 0.04 per cent while oxygen has risen from 0 to 21 per cent.

This transition occurred approximately halfway through Earth's history. Until then, the development of the Earth had been driven by astronomical (solar and planetary) and volcanic processes. At this time, a fourth process, life, became one of the Earth's driving forces. The first of Earth's atmospheres was the original atmosphere acquired from the solar nebula and the second was the one produced by volcanic eruptions and meteoric bombardment. Life's first transformational act in Earth's story was to create the third kind of atmosphere, the one that persists today. One outcome was to freeze the Earth into a snowball.

At 2.5 billion years ago, the Proterozoic Eon had begun (the name means 'early life'). Plants and, possibly, fungi appeared at this time. The upper reaches of the atmosphere formed an ozone layer that protected life on the surface of the Earth from ultraviolet radiation. At the end of the eon the first animals appeared. However, the development of life was held back by a series of global ice ages lasting from 2.45 billion to 2.22 billion years ago, even though the Sun had brightened and the Earth was receiving more solar warmth. The evidence comes from a series of four glacial deposits found in Canada, on the north shore of Lake Huron, as well as other sites in Finland, South Africa, Australia, Antarctica and elsewhere. These are the earliest ice ages so far identified and are known as the Huronian glaciation. Its global distribution suggests the Earth was frozen from the poles to the Equator, a phenomenon known as 'Snowball Earth'.

This dramatic change of climate was triggered by the transition to an oxygenated atmosphere and ocean. The oxygen decreased methane levels in the atmosphere and increased carbon dioxide. Both are greenhouse gases but methane is much more effective than carbon dioxide. Reduction of the blanketing effect of atmospheric greenhouse gases caused the Earth to cool dramatically. Having combined with other gases, the concentration of oxygen in the atmosphere and the oceans reduced. This altered the balance between organisms that rely on using oxygen to metabolize and those that do not. Evolution hedged its bets by pairing organisms into a single organism that metabolized in two ways: one type, like cyanobacteria, metabolized through photosynthesis, the other used its waste products. Single-celled prokaryotes combined together into multicellular forms of life called eukaryotes.

This breakthrough formed the basis for much more complex organisms than those made of a single cell. Cells could readily develop into specialized forms that had different functions that nevertheless operated together to produce a very efficient fit to an environment. However, it took some time for this advantage to take off to full effect – the period from about 2 billion to 1 billion years ago shows few changes in species, geology or climate and therefore provided weak environmental stimuli to provoke evolution; it has been characterized in a term coined by English palaeontologist Martin Brasier as the 'Boring Billion years', the dullest time in Earth's history.

The Proterozoic Eon culminated dramatically in further Snowball Earth episodes, which took place towards the end of the period at about 716 million and 635 million years ago, and by the appearance of new forms of life, grouped together as Ediacaran fauna. Some were large and mobile, with muscular and neural cells, but no skeleton – walking mattresses. Others were tiny worms (like grains of rice). The fossils of one species, *Ikaria,* dating from 550 million years ago, have been found near Nilpena in South Australia. It burrowed in well-oxygenated sand in search of food, and was the earliest-known bilaterian, an organism with a front and back, two symmetrical left and right sides, and openings at either end connected by a gut, so it processed food in a production line. *Ikaria* was an ancestor of most of what we ordinarily recognize as the animal species of today, including ourselves.

The Phanerozoic Eon: life dominates the history of the Earth

The Phanerozoic Eon is the current eon on Earth and started 540 million years ago. The name is derived from the Greek words *phaneros* and *zoe*, meaning 'visible life', and refers to the sudden appearance of an abundance of readily identifiable fossils. It took approximately half a billion years for life to begin on Earth after its formation, but it remained very simple in its forms for a further 3.5 billion years. Once life became complex enough to be visibly modern, it took just another half a billion years for *Homo sapiens* to develop, as we shall see.

The evolution of 'visible life' is something that needs first an organism to arise that is rather simple but then there is a rare combination of circumstances from which carbon atoms can combine in remarkable, complicated and self-replicating molecules and structures. It is a chain of events and developments that stretches from the energy-generating processes in stars that produced the carbon in the first place, through the provision of suitable life-favouring environments on planets, as a by-product of star formation, through the unique chemistry of carbon atoms and the positive feedback mechanisms of evolution to the variety of life on Earth today.

Earth's history is just one example of the way that intelligent life develops on a planet, but it is only one example and we should be wary of reading too much from it. Nevertheless, it appears that it is relatively quick and easy for life to start on a planet in a simple form, but it takes longer and it is more difficult for life to take the first steps into intelligence. One inference is that life might well be found on many planets in our Galaxy but there will be many times fewer that harbour extraterrestrial beings that we can talk to.

The life that first began on Earth in this eon includes sponges, jellyfish, corals, flatworms, molluscs, worms, insects, echinoderms (animals like starfish) and chordates (animals with a spinal column, like us). They evolved into a multitude of life forms in the event referred to as the Cambrian Explosion, a sudden proliferation of organisms and species. The fossils that established this transformation were found in rocks of the Cambrian period, named in 1853 by the Cambridge geologist Adam Sedgwick after Cambria (the Latin name for Wales), where rocks of this time are exposed in abundance.

XI The Crab Nebula and its pulsar in a composite photograph constructed from X-rays observed with the Chandra telescope (blue and white), light with the Hubble Space Telescope (purple) and infrared with the Spitzer Space Telescope (pink). A spinning neutron star formed by a supernova explosion in 1054 CE generates a bright ring of high-energy particles, which leak away and probe the tangles of the nebula's magnetic field.

XII Exoplanet PDS 70b is the bright spot carving a path through the primordial disc of gas and dust around the very young star PDS 70, hidden in this infrared picture behind a central obstruction so that its radiation does not swamp the faint, dusty disc. The planet is more massive than Jupiter and is located roughly 3 billion kilometres from the central star, equivalent to the distance from Uranus to the Sun.

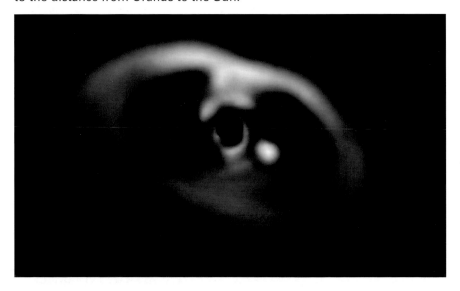

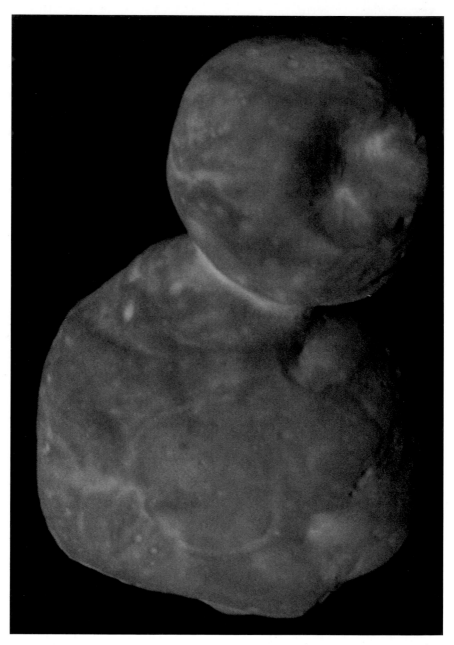

XIII Arrokoth. Orbiting the Sun at a trans-Neptunian distance, Arrokoth is a relic, 36 kilometres (22 miles) long, from the era of planet formation. Its two lobes suggest that it originated as two planetesimals that touched gently and fused at the narrow neck. Its surface is pitted with troughs and craters, some formed from impacts, others from collapsing hollows evacuated by outgassing of volatile ices.

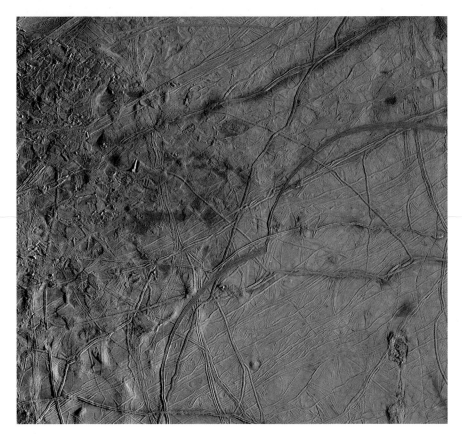

XIV Europa. The icy surface of Jupiter's moon Europa is covered with ridges, bands, blocks and domes (on the left of this image). A flatter plain on the right is marked by arcs joined in cusps, which are fractures induced by tidal stresses caused by Jupiter's gravity on the ice layer that floats on Europa's sub-surface ocean.

XV Gale crater on Mars. NASA's Curiosity rover looked over a long ridge across an undulating plain rich in clay minerals, with light-toned cliffs behind. Haematite, an iron oxide, and the clay minerals in the rocks suggest how the weather changed on Mars, from a watery past to the present, drier, desert climate.

XVIa The Milky Way and the Andromeda Galaxy colliding. Images imagined from the Earth during the collision start with the Andromeda spiral galaxy far off, as now, in a line of sight that passes alongside the Milky Way.

XVIb Viewed above a mountainous terrestrial horizon 2 billion years from now, the separation of the two galaxies has reduced and, appreciably nearer the Milky Way Galaxy, the Andromeda Galaxy has increased in apparent size.

XVIc By 3.75 billion years in the future, the Andromeda Galaxy looms large in the sky as it nears the periphery of the Milky Way Galaxy. The pull of each galaxy on the other has begun to distort their shapes.

XVId The approach is accelerating at 3.85 billion years from now, and the Andromeda Galaxy fills the sky, so close to the Milky Way Galaxy that its gravitational pull has begun to trigger the formation of new stars from nearby gas clouds.

XVIe By 3.9 billion years from now, the sky is filled with new stars that shine with blue light and excite interstellar gas into its characteristic red emissions. Massive stars frequently explode as supernovae, perhaps one per year.

XVIf At 4.0 billion years from now, each galaxy has distorted the other and their original spiral structures have twisted into writhing shapes, clutching each other in a tangled embrace like lizards fighting, throwing the Sun and Earth into a distant orbit.

XVIg At 5.1 billion years, as viewed from the Earth, assumed here to have survived on the periphery of the action, the bright central cores of the Milky Way and the Andromeda Galaxies appear as two bright lobes, side by side.

XVIh By 7 billion years from now, the two galaxies have merged and form a huge and homogeneous elliptical galaxy. Viewed from the distant Earth, the bright core of the elliptical galaxy dominates the night-time sky but star formation has faded away.

XVII Earthrise. While orbiting the Moon in Apollo 8 in 1968, astronaut Bill Anders photographed the Earth, looking beautiful but small and vulnerable, alone in space above the rocky lunar surface.

The Phanerozoic Eon began with life living in the oceans, then some animal species evolved from the shoreline onto dry land in response to the stimuli of varied and ever-changing environments as the tide went in and out. Unicellular plants had already moved onto land 1 billion years or more ago, but evolved into a more familiar plant form about 430 million years ago. At first, plants reproduced by distributing spores, which limited their distribution to swampy land because spores need moisture for the gametes to swim to meet together. Reproduction by seeds enabled trees to spread to make forests on dry land. Trees anchored themselves by growing strong roots into the Earth, stabilizing the land, and became tall by standing on woody trunks by 360 million years ago. This is the start of the geological period called Carboniferous because abundant forests laid down strata of carbonaceous material that became peat and then coal.

Corals are marine invertebrates that first appeared in the sea about 570 million years ago. They built reefs that became sedimentary strata, sometimes of great size, like the present-day Great Barrier Reef in the Coral Sea off Australia, the largest living structure on Earth today, although only one coral reef of many. Other sea creatures deposited skeletal minerals that became rock strata of carbonaceous limestone and chalk, associated with strata of siliceous chert (flint). Life was creating new forms of rock, changing the very make-up and appearance of the Earth's crust in a major way for the first time.

The processes by which life drives forward the history of the Earth at the present time can be typified in the carbon cycle. Carbon is cycled through the atmosphere, rivers, lakes and oceans, and the crust. Carbon dioxide from the atmosphere dissolves in water, forming carbonic acid. This acid combines with calcium and other similar elements to make bicarbonate and carbonate chemicals. These compounds are taken up by molluscs for their shells and fall to the seabed when the molluscs die. They are buried in rock strata through which magma bursts from volcanic eruptions, releasing carbon dioxide through volcanic outgassing. Carbon dioxide is also exchanged from the atmosphere to the crust and back via the photosynthesis of plants and the breathing of animals. Dead organisms ferment and release carbon dioxide and methane, or may become fossil fuels like coal or oil. Human beings speed the fuels to return carbon dioxide into the atmosphere through industrial processes that have newly emerged in Earth's history, starting three hundred years ago in the Industrial Revolution.

In modern times, anthropogenic processes are on such a scale and so quick in their effects that they produce disturbances in the balance of biological cycles, which can be described under the general heading of 'climate change'. The Gaia hypothesis originated by the English chemist James Lovelock in the 1970s suggests that in the longer term the planet will return to its equilibrium, proposing that living organisms interact with the environment so that the whole system self-regulates to maintain favourable conditions for life on Earth. It is a visionary, inspiring and hopeful perspective of environmental science, which, however, remains controversial as an unproved proposal.

Pangaea

Although what happened in the history of the continents in the Archaean and Proterozoic eons is obscure, the evidence from the Phanerozoic Eon of the past few hundred million years is more clearly written. The rocks of the Pangaean supercontinent survive, although they have been mobile. This was the continent that the German geophysicist and polar explorer Alfred Wegener (1880–1930) identified, starting in 1912 when formulating his theory of continental drift, the precursor to the present-day theory of plate tectonics. Wegener noticed that the present-day continents, subcontinents and large islands fit together like the dispersed pieces of a jigsaw puzzle. The east coast of the Americas fits closely onto the west coast of Africa and Europe. Antarctica, Australia, India and Madagascar fit along the east coast of southern Africa. The fit is even better if the boundaries of the continents are identified, not as the present shoreline but as the edges of continental shelves at a depth of 200 metres (650 feet) below present sea level. If the jigsaw is reassembled, the pieces form a single supercontinent, with geological characteristics, including fossil content of the constituent rocks joining across what are now the edges of the pieces. The individual continents separated and then drifted apart at rates of centimetres per year. Wegener's theory was scorned for half a century, but in the 1950s and 1960s further supporting evidence emerged in the form of magnetic data preserved in rocks that showed the same continuity as the fossils.

Pangaea was formed about 330 million years ago by the amalgamation of two earlier large land masses: Gondwanaland, which comprised present-day South America, Africa, Antarctica, Australia, the Indian subcontinent and Arabia; and Laurasia, which comprised present-day Europe (without the Balkans), Asia (without India) and North America. Due to the formation of rifts in Pangaea, it broke up progressively into the present-day continents between 175 million and 60 million years ago. However, the collisions and breakages of the present continents continue even today. For example, the Indian subcontinent is colliding with Asia, forcing up the Himalayas. Incipient fractures at rift lines have appeared at the Red Sea and in East Africa.

The history of land-based and flying animals coincides approximately with the existence of Pangaea. They appeared early in the Phanerozoic Eon: the earliest fossils of millipedes have been found in Scotland from rocks about 430 million years old. Reptiles appeared 312 million years ago, dinosaurs 240 million years ago, mammals about 210 million years ago and birds 150 million years ago. Recognizably modern animals – including humans – evolved rapidly during the last 50–100 million years of this eon. Pangaea straddled the Equator and so it spanned a wide range of climates and environments, in which numerous species evolved. At various times, environments separated and became isolated from or joined to others while drifting considerable distances from one climate system to another. Changes in inter-species competition and in environmental pressures drove species to evolve in different ways.

The evolution of species and therefore the distribution of their fossil remains are inextricably linked with the complex history of Pangaea.

The Chicxulub asteroid strikes

There were five mass extinctions during the Phanerozoic Eon, when large numbers of species died out, allowing new ones quickly to emerge. The causes of all of the extinction events are uncertain, and of some are very unclear. However, they all represent abrupt, globally widespread changes in climate, triggered by prolonged ice ages, extensive volcanic eruptions or violent meteoric impacts. The most widely known and perhaps best understood is the Cretaceous–Palaeogene extinction (previously known as the Cretaceous–Tertiary extinction). The name refers to the boundary between the Cretaceous and the Palaeogene (or Tertiary) geological periods and is abbreviated as K–Pg or K–T, with the K standing for the German word *Kreide*, meaning 'chalk', the rock distinctive of the time. The change of climate represented by the change of rocks above and below the boundary was caused by the impact of a large asteroid in the region now known as the Yucatán Peninsula of Mexico, which struck near the present fishing harbour of Chicxulub (pronounced as 'cheek-shoe-lube').

The Chicxulub event was one of nearly two hundred meteor impacts on Earth, which made craters that survive and whose nature has been confirmed by scientific evidence – there are something like one hundred further credible cases. The craters range up to 300 kilometres (200 miles) in diameter and back to 2 billion years in age. The Chicxulub crater is the second-largest meteor crater known on Earth, but there is very little trace of it readily visible. Glen Penfield, an American geophysicist working for an oil company, discovered it in 1978 in an airborne magnetic survey, which picked up magnetic anomalies that showed as a curious circular arc in the seabed north of the plain of agave plantations and bush near Chicxulub. There is little to be seen on the surface of the land except a shallow trough, perhaps now only

3 metres (10 feet) deep (because the central bowl of the crater has been filled by sediment), and an arc of sink holes (*cenotes* in Mexican Spanish) that continues from the land into the sea. The sink holes are features in the limestone mineral of the area, made by water seeping through cracks made by the impact. The arc on land marks the southern rim of the crater.

Penfield was joined in his investigations of the nature of the circular feature by a PhD student in planetary sciences, Alan Hildebrand. Together, they confirmed that it was a remnant meteor crater by the discovery in the early 1990s of quartz in some sample cores that had been drilled from the area. The shock of the impact had transformed the quartz into minerals like coesite. Coesite is derived from silica and has a dense, heavy structure akin to glass (it was named after American industrial chemist Loring Coes Jr, who synthesized it in 1953 by subjecting quartz to extremely high pressures and temperatures to change its crystal structure). Coesite has been found in craters left by tests of nuclear explosions but had never been found in any naturally occurring rock until it was discovered in 1960 in the Barringer Meteor Crater in Arizona by geologists Edward Chao and Eugene Shoemaker. Its presence is one of the characteristics used to distinguish a meteor crater on Earth from any other kind of hole in the ground.

The Chicxulub asteroid, 10–15 kilometres (6–9 miles) in diameter, plunged through a shallow sea, perhaps 100 metres (330 feet) deep. It impacted into the seabed and shattered, taking a few seconds to start a tsunami and then pulverize and melt the rock of the seabed, excavating in minutes a crater 150 kilometres (90 miles) in diameter and 30 kilometres (18 miles) in depth. Like some lunar craters it has a central hill or mountain, covered since impact by accumulation of ocean sediments in the crater.

Ejected from the crater in a tower of hot gases, superheated steam and glowing, molten rock were 300,000 cubic kilometres (72,000 cubic miles) of rock fragments, which were thrown far away, even into orbit around the Earth. The debris covered the world and its traces are still identifiable as a geological layer in the

Earth's rocks, which contains a high concentration of the element iridium. Iridium, deposited by asteroids on Earth when the planet was formed, is a siderophile and the iridium with which the Earth was born has, mostly, sunk with iron into the Earth's core. Iridium-rich material in the crust of the Earth, such as that in the K-Pg boundary, must have arrived after the formation of the Earth's core. Deep sea drillings showed in 2020 that the thin iridium layer closely overlies the Chicxulub crater immediately underneath the subsequent accumulation of ocean sediments and proves the rapid sequence of events, as well as identifying the nature of the meteor as an asteroid, rather than, say, a comet.

Finely powdered debris remained suspended in the atmosphere for weeks to years, including sulphates from powdered gypsum, the mineral of the Yucatán seabed. It was augmented by volcanic ash from massive eruptions in the Deccan Trap volcanic formation in India, which lasted 30,000 years and spanned the time of the Chicxulub impact. Together, all this powdered material blocked out the Sun, much as would happen after a wide exchange of nuclear weapons in a nuclear war, so that a 'nuclear winter' followed the firestorm of the meteor impact. For a period of time, our blue planet turned white and grey.

These events of 64 million years ago caused a widespread extinction of many land-dwelling species. Most dinosaurs became extinct, although feathered dinosaurs survived, including some that evolved to become birds.

The appearance and development of the *Homo* genus

Small, burrowing mammals also survived and moved to fill the gap in the environment left by the land-dwelling dinosaurs. Prominent among the mammals that eventually developed were Hominoidea (apes, originating 24 million years ago). The evolutionary line that developed from apes towards humankind branched off successively as gibbons, orang-utans, gorillas and chimpanzees, and then hominids such as *Australopithecus*, who originated about 4 million years ago. These hominids developed stone tools and split into two parallel evolutionary branches, descendants of *Australopithecus* and a new genus, *Homo*.

At first, *Australopithecus* and *Homo* coexisted in Africa but *Australopithecus* became extinct, leaving survivors that formed a lineage stretching into *Homo habilis* ('man the toolmaker', 2 million years ago) and *Homo erectus* ('upright man', 1.5 million years ago). Migration spread *H. habilis* and *H. erectus* from Africa into Eurasia and southern Asia. Fossil skulls, partial skeletons and tools of Dmanisi man, found in Dmanisi, Georgia, are 1.8 million years old, the oldest human remains found outside Africa. Human bone fragments of Java man are perhaps 1 million years old. The fossil tibia and flint tools of an individual of *H. erectus* from 700,000 years ago were found in 1982 at Boxgrove, near Chichester in England. Forty individuals of *H. erectus* known as Peking man date to 400,000 years ago. The lineage of *H. erectus* from these migrations appears then to have foundered.

The most recent 'out of Africa' migrations of the genus *Homo* followed a similar migration track to that of *H. erectus* into Eurasia from 300,000 years ago. In these migrations, *H. neanderthalensis* and *H. sapiens* (modern humans) coexisted and indeed interbred. Neanderthals became extinct as a separate species of *Homo* about 40,000 years ago, but some Neanderthal DNA lives on in *H. sapiens*, which survived as the dominant and then sole human species.

For more than a million years of geological history, fossils of the genus *Homo* are associated with increasingly sophisticated shelters, domestic fires, stone tools, engraved shells, carved sculptures, bone flutes and other musical instruments, and cave murals. The murals were often of animals hunted for food and painted silhouettes of human hands, serving early artistic and ritual purposes. Civilization gradually dawned.

From 300,000 years ago, *H. sapiens* spread from Africa into southern Asia, reaching to Australia by 60,000 years ago and back into Europe by 40,000 years ago. The American continents were the most recent to be occupied by humans, with North America reached by 20,000 years ago, either by land travel from Mongolia and the now collapsed Bering land bridge southwards through Canada, or by a combination of sea and land travel to South America and northwards to North America, again over land. *H. sapiens* colonized the entire world (except for Antarctica) and began wreaking changes on the planet and its ecosystem.

At first, humans were nomadic, following migrating animals, and, as in the case of the herds of mammoths that once roamed the North American plains and the Russian steppes, hunting them mercilessly and contributing to their extinction. This started the Anthropocene epoch in geology, in which specifically human life is a distinct force in the history of the Earth (the first part of 'Anthropocene' comes from the Greek word for 'human'). An early trace of this era can be seen in the dense archaeological stratum of mammoth bones at the killing field in Clovis, New Mexico, one of the earliest surviving scenes of human activity in North America, 13,000 years ago.

Gradually, humans switched their mode of living to become sedentary. They initiated agriculture, clearing forests to grow crops, replacing virgin forest with cultivated landscapes. Humans tamed rivers and lived in permanent settlements. From 9,000 years ago, humans built cities, adding man-made structures of a geological scale to the landscape. Within the last half-millennium, industrial activity began to alter the very composition of the Earth's surface and its atmosphere. Humans changed geology, altering the shape of the land through large-scale activities of mining and civil engineering and the disposition and mix of its vegetation through land clearance and agriculture. With the population of *H. sapiens* exploding, human pressure on the habitats of other species has been leading to their decline and, in too many cases, extinction en masse.

If we represent the biography of the Universe as a timeline that is as long as all the writing in this book, strung out into a single row, the Earth was born somewhere in the middle of the words of Chapter 7. *Homo sapiens* occupies the book's last word or two, and human civilization less than the width of the last letter. Your lifetime, as part of that cosmic history up to the present moment, is represented by a small fraction of the last full stop.

12

Sequel: The Future Life of the Universe

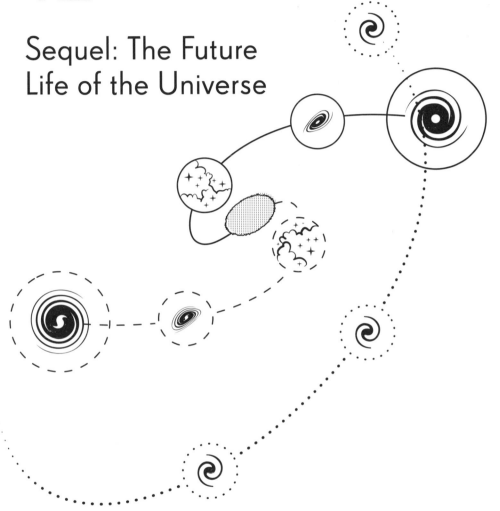

This book is a biography of the Universe and mostly looks
backwards, but a peek into its future life shows where we,
our planet, our Sun and our Galaxy are heading.

The Andromeda Galaxy (upper right) and ours (left) are approaching one
other, each distorting and triggering the formation of stars in the other.
They will collide in 4.6 billion years, merging to form a single, giant elliptical
galaxy. The Triangulum Galaxy (top right), the bystander in the threesome,
will curve in a trajectory past the two of them.

The near future on Earth

The same large effects that have occurred from natural astronomical phenomena over the past millions of years of the Earth's life are also likely to occur in the future. Planet-wide volcanic eruptions and large meteor strikes at the scale of the events that have already occurred and provoked the mass extinctions of geological history remain possible, likely or even inevitable, but unpredictable. In addition, as described in the previous chapter, human beings have developed to become a powerful planetary force and we are already making a mark on other worlds through space exploration, leaving footprints and wheel tracks on the Moon, and abandoned equipment on many worlds. We may well imminently become a potent interplanetary force, the ultimate scale of which is the proposal to 'terraform' Mars, thus far science fiction but potentially a real engineering project. It would be intended to change Mars from a desert into a verdant, habitable planet (let us hope that the project is well thought through and has the desired effect). We may have to do something similar to Earth to reverse the effects of climate change – for example, to remove carbon dioxide from the atmosphere on a global scale and store it in the ground.

However, given that we are an animal species, history suggests that eventually our species will become extinct, presumably on a timescale comparable to the period during which we have evolved in the past. Our species has lasted a quarter of a million years and our distant ape-like ancestors lived only millions of years ago. If the evolutionary forces remain natural ones in the future, this suggests that, in a similar time, we will have descendants but they will be as different from us as we are from our ancestors. There are those who argue, however, that this transformation of our species and our environment will happen much more quickly, since changes are taking place at an accelerating pace, driven by rising population numbers and increasing technological capability.

The range of possible ways in which humans might self-inflict extinction is wide, including not least:

- widespread starvation and thirst due to unfulfilled demands by the increasing human population on agriculture and natural resources;
- pandemics like Ebola, influenza and Covid-19, caused by malpractices in animal husbandry and the crossing of viruses and other pathogens from wild animal and bird populations to human beings;
- release (accidental or otherwise) of toxic industrial products, such as the CFC chemicals that, until banned from use in refrigerators, depleted the ozone layer when they escaped;
- all-out nuclear war; and
- anthropogenic generation of carbon dioxide and methane that will, if unchecked, cause damaging global climate change.

The last-mentioned climate catastrophe is widely described as imminent: it is a matter for debate whether we will be able to generate the political will that can make changes on a scale and with a speed to control anthropogenic climate change. It seems certain that the Earth's atmosphere will not be able to respond enough and quickly enough to mitigate the immediate effects of human activity, although there is some hope in James Lovelock's Gaia hypothesis that life will, on a longer timescale, be able to adjust the environment to maintain itself (see page 242), although that may not favour humans.

Our species may thus become extinct rapidly, or at least have its capabilities reduced enough for humans to cease to dominate environmental changes. If so, the present Anthropocene epoch will be quickly over. In retrospect, as viewed by surviving descendants, it will be marked by the significant changes in the landscape that human beings made to favour their own activities like agriculture, mining and transportation, and the control of flowing water. Additional changes will have been made by the unexpected consequences of human activity, such as alterations of the coastline that will have shrunk the area of dry land as a result of the rise of sea level, itself the result of the increase in the amount of carbon dioxide in the atmosphere.

The geological characteristics of the epoch will include rather thin but distinctive layers in the Earth's geological record. The strata associated with the Anthropocene will comprise human-made scrap, such as building rubble, worked metal and long-lasting plastic. Other components of the strata will include carbon ash from energy generation from fossil fuel, radioactive elements from nuclear bomb tests and the waste and accidental discharges from nuclear energy generation and from medicine, and, eventually, the decay products of plastic, both in localized earth-based deposits and in more widely dispersed, thin layers in sedimentary deposits at the bottom of the ocean.

When might this be? The Doomsday Clock is published by the *Bulletin of the Atomic Scientists*. It is an image of a clock-face that is ticking towards a symbolic midnight at the future moment of human-made global catastrophe. It is a concept that warns about how close we are to destroying our world with dangerous technologies of our own making. The metaphor uses the imagery of apocalypse (midnight) and the idiom of nuclear tests (countdown to zero). It is a reminder of the urgent perils that we must address if we are to survive on our planet.

The original setting of the Doomsday Clock in 1947 was seven minutes to midnight because of the threat from use of nuclear weapons. With the end of the Cold War, it was put back to seventeen minutes to midnight in 1991, its most optimistic setting ever. In 2007, the Clock's adjustments began to be influenced by non-nuclear threats, particularly the threat of a climate catastrophe, and it has persistently neared the critical hour. In January 2020, the Doomsday Clock was advanced to one hundred seconds to midnight, the closest that it has ever been. It is maddening that the Clock is not (cannot be) calibrated, so we do not know how much time this really represents, and of course the purpose of the Clock is to prompt us to act to avert catastrophe, so that midnight never strikes. However, the symbolism suggests that we are worryingly close to the critical point.

The end of total solar eclipses

On a less worrying note, we can certainly predict the termination of one of astronomy's greatest spectacles – total eclipses – because the Moon is getting further away. The Moon moves in an elliptical orbit, so its distance varies somewhat through a month, but it is at an average distance from the centre of the Earth of 384,400 kilometres (238,855 miles), which is 1.28 light seconds. A laser pulse transmitted from a telescope on Earth, reflected from mirrors left on the Moon by Apollo astronauts, and returned back along the same path to a detector in the same telescope, takes 2.56 seconds for the return journey. Measuring that travel-time accurately is how astronomers monitor the orbit of the Moon.

The Moon is spiralling outwards in its orbit. Its drift deeper into space has been happening for a long time, ever since its creation more than 4 billion years ago (see Chapter 10). This has been directly confirmed for the last half a billion years by analysis of the Moon's effect on oceanic tides, which leave their traces in geology. Marine organisms such as coral, bivalves, brachiopods, cephalopods and stromatolites feed in tidal waters and show growth rings in which it is possible to see daily, monthly and annual periodicities, linked respectively to the Earth's rotation, the Moon's orbit around the Earth and the Earth's orbit around the Sun. Analysis of fossils from 70 million years ago shows that, in the Late Cretaceous period, there were 372 days per year, so that the day was then about 23.5 hours long. Sedimentary rocks are likewise laid down with tidal rhythms, and that has made it possible to estimate that towards the beginning of the Cambrian period, 620 million years ago, the day was 22 hours long. Since that long ago, the Moon has retreated at an average rate of about 2.2 centimetres (0.9 inch) per year, about the same speed that fingernails grow.

At the present time, the Moon has accelerated to retreat at 3.8 centimetres (1.5 inches) per year and the 'day' has lengthened to 24 hours. The acceleration is thought to have happened because the Moon's orbital energy is being dissipated more quickly. Continental drift, volcanic activity and geological forces in general have created seas that are the right size and shape to be resonant to the oscillation of seawater. The seas suck up energy in the tides, draining it from the Moon's orbit.

It is a strange coincidence that the Sun and the Moon are the same size in the sky. The Sun is four hundred times the diameter of the Moon but it is four hundred times further away. From time to time the Moon passes exactly over the Sun, and we can experience the glory of a total solar eclipse. At other times the Moon is a bit further away than usual and, although the Moon and Sun line up, the Moon leaves a ring of sunlight around its edges – this makes an annular eclipse. This is not so glorious as a total eclipse because the sunlight left peeking over the Moon's edge overwhelms the faint light from the Sun's corona and prominences. The darkness on Earth as sunlight is completely cut off, the sudden revelation of the faint and streaky white corona and the beautiful scarlet prominences give a total eclipse its impact. By contrast, annular eclipses are interesting but underwhelming.

Because the distance of the Moon is increasing, annular eclipses are becoming more common and total eclipses are becoming fewer, at a slow rate of change. Eventually the Moon will be so far away that it will no longer be the same apparent size as the Sun, but will always fit inside the Sun's shape so there will only be annular solar eclipses. The demise of total eclipses will not occur suddenly, because the Moon's distance from Earth varies so much, but total eclipses will be progressively less frequent over the next 400 million years. As the Moon's distance and other characteristics of its orbit change, total eclipses will be on-again, off-again for a further 400 million years. If, in spite of the earlier discussion about our impact on the planet, Earth is still inhabited, people living after then will have permanently lost one of the greatest astronomical spectacles.

The distant future of the Sun and Earth

The Earth's surface will change its character when plate tectonics cease to push the continents about. This will happen when the outer layers of the Earth have cooled enough to completely solidify, perhaps in a couple of billion years. This will be the end of the mountain-building era on Earth, apart from the walls of craters made by occasional impacts of large asteroids. Mountain ranges will gradually wear away under the processes of erosion, becoming hilly plateaux. Individual volcanoes or clusters of them may grow for a while, building over weak spots in the crust, as they do in Hawaii and on Mars and Venus, two planets that, like the Earth in this future, do not have tectonic plates.

Even this reduced volcanic activity will cease as the Earth cools further. Earth will begin to die, with no more seismic activity from earthquakes or volcanoes. Eventually, its liquid iron core will solidify and its convection seize up. Our planet's magnetic field will die away completely and permanently. Unlike the temporary loss of the geophysical magnetic field during the switches in its polarity (see page 227), this permanent loss will be catastrophic. Unimpeded, the Sun's particles will eventually scour away the atmosphere. With no air pressure to keep water molecules from escaping from seawater, the oceans will vaporize, rainfall will cease and the land will dry out. In 1 billion to 2 billion years from now, the Earth will have lost its equanimity and become a quiescent desert land. Earth will have turned into Mars.

The present-day physics of the Sun, its structure and the nuclear energy that it generates are very well understood. Predictions of the Sun's future based on the same physics have been made and published repeatedly. However, during its evolution the Sun grows in size, its surface gravity reduces and it loses mass easily. What happens depends critically on the way mass loss happens and this is not a very secure area of astronomy. In 2008, University of Sussex astronomers Klaus-Peter Schröder, Robert Connon Smith and Kevin Apps revisited the 'Distant future of the Sun and Earth' paying special attention to the mass loss. Their predictions started off in a familiar way but then started to differ significantly from what had been conjectured before.

Over the near future, the Sun will warm and brighten, increasing in luminosity by 1 per cent every 110 million years. In 1 billion years, it will be about 10 per cent brighter. The oceans will start to evaporate and the water vapour content of Earth's atmosphere will increase substantially. Water vapour is a greenhouse gas, so runaway evaporation will cause the surface to become unbearably hot and the oceans to boil dry. In the stratosphere, solar ultraviolet light will break up the water molecules, which will gradually escape into space.

At the same time that the Earth's magnetic field collapses, the Sun will be much hotter. It will be at its hottest 2.55 billion years from now, and at its brightest as a hydrogen-burning star 5.4 billion years from now, emitting nearly twice the energy that it did originally. This will put the Earth in the same position then that Venus is in now, so far as the flux of solar energy that Earth receives. This will cause the Earth to bake, at about the same time that it loses its magnetospheric defence against solar cosmic radiation. It is hard to see that life could continue on our planet. At the same time, however, the more distant parts of the solar system will also become warmer. Satellites like Jupiter's Europa, where water is abundant but is currently solid ice, will become more habitable and those like Saturn's Titan, where organic molecules abound, may feel the push in the Sun's radiation to develop life. These satellites may turn into the Earth as it was 3 billion years ago in the Archaean Eon.

In 5.4 billion years, the Sun's core will run out of hydrogen fuel and start burning helium. The Sun will both expand and cool as it becomes a red giant star. Its radius will grow as much as 250 times, and its rotation will slow from a period of a month to thousands of years. It will lose mass and that will decrease the gravitational attraction of the Sun for the Earth and cause the Earth's orbit to move outwards. Earth will just keep ahead of the Sun's surface as it expands to become a red giant star. Mercury and Venus will be swallowed, but not the Earth, although it will feel the effect of the Sun's atmosphere.

The Sun will remain a red giant for 2 billion years. Its external layers will dissipate into space and altogether, its mass will diminish to about half what it is now. The tidal forces between the Earth and the Sun and friction of the Earth in its orbit through the Sun's atmosphere will cause the Earth to spiral into the Sun in about 7.6 billion years in the future, unless something happens, like a close encounter with another star, to strip Earth out of its orbit.

Because of the loss of mass, the core of the Sun 7.7 billion years from now will become exposed. It will light up the mass that the Sun lost into the surrounding space. The Sun will be the central star for a planetary nebula. Some planetary nebulae are very spectacular but the Sun's will be puny. It will be similar to the unspectacular IC 2149, a planetary nebula that also evolved from a star of about 1 solar mass. The planetary nebula will dissipate and the Sun will go on to become an unremarkable white dwarf, which fades and becomes even more unremarkable.

It may well be that the white dwarf will drag in crumbled remains of the Earth, if any of it survives. Terrestrial rocks will salt the white dwarf's atmosphere and vaporize. As a proof that this actually happens sometimes, the remnants of exoplanets have been detected as chemical pollutants in the atmospheres of many white dwarfs. In one case, planetary fragments from a planet nearly the size of the Earth have fallen into and polluted the atmosphere of the white dwarf GD 61. If this happened in one event, it happened quickly, within the last hundred years, because this is the length of time that it takes for heavy elements to sink into the white dwarf's atmosphere to such a depth as to become invisible. It was an event that foreshadows the likely fate of our own planet. If the Earth makes it to this stage of the Sun's and the solar system's evolution, the last evidence that the Earth ever existed will be detectable only for brief centuries as minor chemical constituents in the atmosphere of a fading white dwarf star.

Baked, dried, crumbled, engulfed, vaporized in the atmosphere of a fading star: the likely future of our fragile planet (pl. xvii) is a progressive path through successive stages of destruction on the way to oblivion.

The future of our Galaxy

In 1913, American astronomer Vesto Melvin Slipher (1875–1969) measured the speed of the Andromeda Galaxy along the line of sight – it was 310 kilometres (190 miles) per second relative to the Sun, the highest speed of anything known up to that time – and was coming towards us. When in 1929 Edwin Hubble examined the speed of forty-six galaxies as part of his discovery of the expanding Universe, it became clear that such a high speed was not at all unusual, but the Andromeda Galaxy was unusual in that it was approaching us, the only galaxy of the forty-six that was doing so. Galaxies' speeds are proportional to their distances (Hubble's Law; see page 20), so in general the further away, the faster they are receding, but there is some scatter in this relationship. Some of this scatter is caused by the motion of the Sun in orbit around our Galaxy, but even when this is taken into account, the Andromeda Galaxy is approaching our Galaxy. Analysing this situation in 1987, British astronomer James Binney and Canadian astronomer Scott Tremaine suggested that our Milky Way Galaxy will collide and merge with the Andromeda Galaxy in about 2 billion to 5 billion years.

There was a large uncertainty on Binney and Tremaine's calculation, because, although it has been possible for a hundred years to determine the radial component of Andromeda's velocity, it has been very difficult indeed to determine the speed of any galaxy across the line of sight because galaxies are so far away that angular displacements due to that motion over the duration of astronomical observations (decades to centuries) are tiny. This changed with the launch first of the Hubble Space Telescope, then of the Gaia space satellite (see page 100–102), which are able to measure such velocities to unprecedented accuracy, even during the few years' lifetime of a space mission.

A large group of scientists led from the USA by Dutch astronomer Roeland P. van der Marel not only were able to determine from Gaia data in 2019 the fact that the Andromeda Galaxy is indeed falling towards the Milky Way at a speed of about 130 kilometres

(80 miles) per second, but also were able to calculate some of the details of the future encounter (pl. XVI). Starting from the mass of the two galaxies, and the distance that they are apart, and their relative motion, the astronomers calculated their future trajectory – with some uncertainty because all those 'givens' are imprecise. The two galaxies will collide in 4.6 billion years, not head-on but with more of a glancing blow. The central regions of the Milky Way will pass through the disc of the Andromeda Galaxy. The team described what happens in the collision by modelling the distribution of stars in both galaxies and seeing what happens to them.

The arms of the galaxies will twist and fly outwards as they make their closest approach. Because the galaxies are effectively getting bigger, their energy of motion is fed into the motion of their constituent stars, so as a whole the two galaxies lose speed and on the outward journey after closest approach they will not separate by a large distance, but quickly come to a halt and fall back together. Over the following billion years they will move in and out a few times in (relatively) quick oscillations. Within the galaxies the stars will not in general collide but will stream through each other like the bands of guardsmen on parade. However, their gas clouds will collide; this will trigger the formation of new stars and there will be a number of bursts of star formation synchronized with the multiple oscillations. This will use up virtually all the hydrogen gas in both galaxies and star formation will cease. The stars of the two galaxies will get muddled up and they will merge, becoming a single elliptical galaxy.

The Milky Way and the Andromeda Galaxy belong to the Local Group of galaxies, with M 33 (the Triangulum Galaxy) being a satellite of Andromeda. It is carried along in the collision but survives in the outer reaches of the new elliptical galaxy.

It is not possible to predict what will happen to an individual star, like the Sun, in the galaxies in the collision. Depending on the exact timing of the collision, the Sun may or may not still have the Earth as one of its planets. In the turmoil of the collision, stars from both galaxies will whizz about in disorder. One may pass close to the Sun and possibly strip the Earth out of its orbit. In any case, the Sun will be a red giant soon after the start of the collision and, if the final years of its maturity are spent zooming out into intergalactic space, the nebula that it generates as it turns into a white dwarf might be something that can be seen from few other stars and planetary systems. The Sun's end might be lonely. But another possibility is that the Sun might end up as a member of the halo of the elliptical galaxy, together with tens or hundreds of billions of similar stars, orbiting quietly in a well-populated celestial rest home. The galaxy's lights will be turned off one by one as the individual stars die, the Sun among them, and become white and then black dwarfs. Aged thousands of billions of years, the merged galaxy will become a galaxy of completely dark stars.

Cosmic expansion

Will our Universe keep expanding forever or will the expansion halt after some time, reverse and become contraction, leading to a final 'big crunch'? The answer to this question is helped by knowing how much matter and energy there are in the Universe. The more there are, the greater the gravitational pull they have on the expansion. (Energy has an equivalent mass, given by Einstein's famous formula $E = mc^2$.) The scientific way to pose this question is through the shape of the Universe: is it open (will expand forever) or closed (will fall back into itself), or flat (balanced between the two)? These are terms from geometry: the connection between geometry and cosmology is through general relativity, which talks about the curvature of space and time caused by the presence of mass.

The geometry of the Universe is measured by the density parameter, which is the ratio of the actual density of the Universe to the density of a flat Universe. If the Universe is flat and contains just the right amount of mass (and energy), the density parameter is exactly 1. If the Universe is open the density parameter is between 0 and 1; if the Universe is closed the density parameter is more than 1. If the density parameter is exactly 1, the Universe is flat and on the boundary between open and closed. The density of the Universe in this dividing case is called the critical density.

One way to go about addressing the issue is to look at how much mass there is in the Universe – for example, by seeing how many stars there are in how many galaxies, or how much hydrogen there is in intergalactic space. These methods of direct detection fail to find anything like the amount of mass needed to close the Universe.

Another method is to see how fast things like galaxies move, so you can tell how much mass is moving them through their cumulative gravitational pull. This method detects all of the matter in the Universe, dark matter as well as ordinary matter. Even with dark matter included, the total amount of matter detected by this method is still far short of the amount needed to close the Universe.

The best method found so far to determine cosmological parameters is to examine the statistics of the fluctuations in the CMB in combination with other key cosmological relationships. The average size of the spots is sensitive to the magnifying effects of the curvature of space: in a flat Universe, there is no magnification; in a closed Universe, the size of the spots appears larger than they really are; in an open Universe, the spots appear smaller. The Planck space satellite spent four years in orbit gathering ever more accurate data on the brightness of the CMB (see Chapter 2); its science team spent five more years analysing the data and concluded in 2019 that the density parameter is 1, to within 3.5 per cent.

The geometry of the Universe is thus flat and it will expand forever, with the galaxies fading and separating further and further apart. The remote past of the Universe before the Big Bang is the metaphorical darkness of ignorance, but the remote future of the Universe is literal darkness, punctuated from time to time only by the unheard explosions of merging supermassive black holes, radiating whisper-quiet gravitational waves.

13

Prequel: What Caused the Big Bang Expansion?

The birth moment of the Universe

When artists want to picture the start of a grand affair, they might identify the particular, briefest instant. The scale of what follows is something for the viewer to imagine, and is thus more mysterious, much greater and longer-lasting than the artist could show. Michelangelo did this in his painting of the Creation on the ceiling of the Sistine Chapel of the Apostolic Palace in Vatican City. He made the pivot on which the decoration turns the gentle moment in which the tips of God's and Adam's forefingers reach to touch, starting the history of humankind and putting in train the Christian story. The momentous birth of the Universe was likewise brief and very localized, but, in contrast to Michelangelo's picture, very, very violent.

After birth, the life of the Universe became the subject matter of astronomy, identifiable in the large-scale and long-lasting history described in this book. The circumstances of the birth itself can be imagined, up to now at least, only as ideas enunciated through theoretical physics. According to these ideas, the Universe originated in the smallest imaginable time, only 10^{-32} of a second (10^{-32} is a number that can be expressed as a fraction by a numeral 1 divided by a number written as a 1 with 32 zeroes after it). The word 'astronomical' has, by metaphor, gained the secondary meaning of 'enormous', but this astronomical number must be one of the smallest in the whole of science, if not the smallest.

The Universe came into existence with an initial density and temperature that were unimaginably high. Under such compressed conditions, the Universe began united into one single system of forces and energies. This is the era in which its physics could only be described by the Grand Unified Theory, the one being sought by people like English theoretical physicist Stephen Hawking (1942–2018), rather than by the theories of general relativity and quantum mechanics separately. The four fundamental forces of nature – electromagnetism, weak nuclear force, strong nuclear force and gravitation – were unified into one fundamental force, expressed in many dimensions beyond the ones that we now perceive. The high density and temperature caused a tremendous pressure so the Universe began to expand and cool, and the forces separated. Gravitation separated from the others, followed by the separation of the strong nuclear force from electromagnetism and the weak nuclear force – this separation is why we now think of these forces as different one from the other. Time took on a different appearance from the three spatial dimensions. Some additional dimensions beyond the four that we now identify shrank and disappeared from view.

The Universe was expanding at about this time in a process termed 'cosmic inflation', a theory developed separately by American theoretical physicist Alan Guth (b. 1947) and Russian-American astrophysicist Andrei Linde (b. 1948). It started with a brief burst of exponentially rapid expansion, called inflation, which led to the more uniform growth that is happening today. During inflation, the Universe expanded by a huge factor – in that initial 10^{-32} seconds it doubled in size eighty times or more – that is, a factor of 10^{26}.

The idea of inflation was the invention of Guth in 1979, which he almost stumbled over while tackling a problem of quantum mechanics and particle physics: namely, why there are no magnetic monopoles. Magnets always come as pairs of poles, a north and a south pole together, never as one or the other. The solution to this problem led him to the thought that, when it was dense and hot, the Universe once had monopoles in abundance but that they have all separated one from the other by a huge expansion of space. A short time later, Guth realized that this rapid inflation provided answers to two puzzling questions about the Big Bang: the 'horizon' problem and the 'flatness' problem.

The horizon problem is the following. The heat left over from the Big Bang is the so-called Cosmic Microwave Background radiation (CMB), which is almost completely uniform across the sky. Opposite parts of the sky, separated by many billions of light years, have never had a chance to interact with each other but nevertheless look the same. How could this be? No physical process could have acted to cause them to be equal.

The flatness problem is the question of why the content and speed of expansion of the Universe are so exactly balanced that it will just exactly expand for ever (see page 260). Inflation gives an answer: if there was some curvature at the outset of inflation, expansion will increase the radius of curvature and flatten the curvature. The Earth has a large radius of curvature, and it looks flat from where we stand on its surface, which is why the Flat Earth theory retained credibility for so long. Likewise, the Universe expanded to such a size that it is now flat.

These two problems were addressed by Guth in a paper entitled 'Inflationary Universe: A Possible Solution to the Horizon and Flatness Problems', published in 1981. Guth suggested that the Universe was originally in a state from which it decayed, expanding and liberating the energy that the Universe has today. The early phases of the expansion exposed different parts of the Universe to one another and that is how they all became the same; they retained their similarity as they separated. This solved the horizon problem. The expansion also ironed out any primordial curvature, making the Universe flat. This solved the flatness problem.

There were some theoretical difficulties with Guth's version of inflation theory, which were addressed in 1982 by Linde. Linde was born to parents who were both physicists and grew up in Moscow. He married a physicist, Renata Kallosh, both of them moving to the United States in 1990 and becoming professors at Stanford University.

A third player in the inflation story was Russian astrophysicist Alexei Starobinsky (b. 1948), who was able to link some generalities of quantum physics with general relativity in describing how the Universe expanded. Starobinsky was able to predict from his model of inflation that the Big Bang would have generated gravitational waves. These gravitational waves would carry a picture of the conditions in the Big Bang as it was at this time, just as the CMB carries a picture of the Universe when it was 380,000 years old.

A gravitational wave detector known as the Laser Interferometer Space Antenna (eLISA – the prefix 'e' distinguishes the current design from a previous version) is being planned by ESA to detect these gravitational waves (as well as gravitational waves from binary stars and merging black holes). eLISA will orbit in space, so it does not have to cope with terrestrially generated motions like Earth tremors or the rumble of heavy traffic on an adjacent motorway, which confuse the operation of ground-based gravitational wave detectors but not one that operates in the peace of space (although it does have to cope with random impacts of material ejected by the Sun). Moreover, eLISA is being built to a gigantic scale, which makes it possible to detect low-frequency gravitational waves. This is a significant band in the gravitational wave spectrum that is thought to be radiated by close binary stars and supermassive binary black holes.

eLISA will consist of three spacecraft positioned at the corners of a triangle in space whose sides are 2.5 million kilometres (1.5 million miles) long, six times the Earth–Moon distance. The centre of the triangle will lie 50 million kilometres (31 million miles) from the Earth, following our planet in its orbit around the Sun. Each spacecraft will fly in a wavy orbit in such a way that the triangle rotates like a cartwheel about its centre and will carry two

46-millimetre (1.8-inch) cubes made of a gold-platinum alloy. The cubes are designed to float free inside a chamber (just as an astronaut floats free inside a space station) and are made of an alloy that is non-magnetic so the cubes do not respond to magnetic fields, with arrangements made to discharge any electrostatic charge that builds up so the cubes are immune to space weather and solar cosmic rays. The spacecraft will act as a shield to stop the solar wind from buffeting the cubes and disturbing their measurements of the much more subtle gravitational waves. The position of each cube is sensed and the spacecraft and its chamber are adjusted to maintain their position around the cube so that the cube is always in free-fall, able to respond solely to gravity. The separations of the cubes along the sides of the triangle will be measured by a laser interferometer system to an accuracy of 20 picometres (one-billionth of a millimetre – the diameter of an atom).

eLISA will detect gravitational waves as they pass through the solar system, causing the cubes to bob back and forth within their chambers. A test spacecraft launched in 2015, LISA Pathfinder, spectacularly proved that the technology proposed for eLISA works, so it is now being built and is scheduled for launch in the 2030s. It will become by far the largest scientific instrument ever made, almost on the scale of the solar system itself. It should be able to detect the Gravitational Wave Background from the inflationary period of the Big Bang. These gravitational waves are messengers from the Big Bang generated by random, independent events combining to create a cosmic gravitational wave background. The individual events are from dense volumes of Big Bang material that encounter each other or which oscillate. They produce random gravitational waves that make a continuous noise (much like radio 'static').

Cosmic inflation is a cosmological idea that, at first hearing, seems whacky. eLISA will tell us if cosmic inflation really means something. It will eventually show us what the start of everything was like, back in that first 10^{-32} of a second.

Quark-gluon plasma

When the temperature of the Big Bang material had reduced to some thousands of billions of degrees, the particles of which it was made had energies that are comparable to the highest energies reached in terrestrial particle accelerators, such as the Large Hadron Collider at the CERN laboratory in Geneva. The properties of fundamental particles established at places like CERN are properties that pertain to the earliest times in the Big Bang for which we have real evidence.

As described in Chapter 2, the Big Bang material right after cosmic inflation was made of a mixture of all the known fundamental particles, right down to the most fundamental particles currently known called quarks and gluons. These constituted the majority of particles at that time in a mixture known as a quark-gluon plasma. Quarks are the building blocks of protons and similar particles; gluons are the particles that carry the strong force that binds the quarks. The plasma also contained leptons, including electrons and muons and neutrinos, and the particles that carry the forces that act between these, including high-energy photons.

As the Universe aged to 1 millisecond, the plasma had cooled enough for triplets of quarks to bind together with gluons to make protons (two up-quarks and a down-quark) and neutrons (two downs and an up) and their antiparticles. Protons and neutrons are called baryons, so this is the time at which baryonic matter – our kind of matter – came into existence.

What about dark matter? Did it come into existence during inflation, before the hot Big Bang, or was it during the Big Bang itself, in parallel with baryonic matter? When cosmologists know what dark matter is, they might be in a better position to say.

The mixture included antiparticles as well as particles, produced in almost equal numbers. Antiparticles are complementary to particles. If a particle meets its antiparticle, they mutually annihilate – the particle falls into a metaphorical hole where it fits exactly. The result is empty space and energy, which radiates away. It is theoretically possible for antiparticle galaxies to exist that are indistinguishable from ordinary galaxies. They would contain anti-suns, anti-planets and alien anti-beings. It would be unsafe for us to shake hands if we encountered such a person – we would together result in mutually assured destruction.

The symmetry in the Big Bang between particles and antiparticles was not precise. For every billion antiquarks, there were a billion and one quarks, so when they all had touched and annihilated each other there was one quark left over, which is why the Universe consists of matter with no antimatter (except on a very local scale for a very short time after a very rare particle physics event of some energetic sort).

The Multiverse

In some versions of cosmological inflation developed by Linde, the Universe consists of separate pockets of exponentially large regions each with its own characteristic physics. The inhabitants of each pocket see themselves as living within an isolated region and think that it is the entire Universe, but there are unseen pockets all around. Such a system is called the Multiverse.

This concept provides an explanation for the anthropic principle, which tries to deal with the problem that the physics of the Universe seems fine-tuned in some ways so that it is possible for us to exist. The most startling example of the anthropic principle was discovered in 1952 by University of Cambridge astrophysicist Fred Hoyle (1915–2001) as a result of studying the 'triple-alpha' nuclear reaction as he searched for the reaction that powers red giant stars. Stars make carbon by combining two helium nuclei to make a beryllium nucleus and adding a third to make a carbon nucleus. Helium nuclei are also known as alpha particles, hence the term 'triple-alpha'. The beryllium nucleus is unstable, so when triple-alpha was put forward as the process by which carbon was made in the Universe, it used to be thought that it would decay before the third alpha particle could combine with it. This would mean that carbon could not be made in stars, but this was evidently not so: carbon was not made in the Big Bang, it could be made only in stars, and it is the fourth-most common element in the Universe, made in abundance.

Hoyle realized that there must be a resonance (an enhanced interaction rate) in the second step of the triple-alpha process that helps the carbon to form before the beryllium disappears. In 1953, Hoyle travelled to the Kellogg Radiation Laboratory at Caltech to ask for help from researchers, including nuclear physicist William Fowler. He found it difficult to persuade the sceptical physicists to look for the resonance, but a relatively junior physicist, Ward Whaling, who had just moved to Caltech and was seeking a project, took up Hoyle's suggestion and within a few months found the resonance. Fowler was so impressed by this result that he turned to investigate with Hoyle and the astronomers Margaret Burbidge and Geoffrey Burbidge how nuclear reactions in stars made all the elements (save hydrogen and helium). Fowler received the Nobel Prize in Physics in 1983 for this work.

The critical resonance that makes the manufacture of carbon in stars possible depends on a coincidence among energy levels in three separate nuclei: helium, beryllium and carbon. The coincidence is so precise that it looks as if the nuclear physics of the three nuclei has been arranged to make carbon abundantly. Since carbon is essential for life to exist, it seems the Universe has been made for our benefit. This is an anthropocentric (meaning 'centred on humankind') point of view, which in medieval times was thought to be literally true of the Universe as well as true metaphorically, with humankind the exclusive focus of God's concern and the resources offered by Nature. Such views have lost much of their force since Copernicus revealed in 1543 that Earth is not the central focus of the orbits of the stars and planets (see page 37).

There are other scientific coincidences with an anthropic effect similar to triple-alpha that may have been organized by a powerful, beneficent being, who created the Universe for us. This is logically possible and is an argument that appeals to theologians because it provides a scientific context for the First Cause.

The First Cause is an argument for the existence of God, otherwise known as the cosmological argument. It originated in works by the Greek philosophers Aristotle and Plato, and in the West is associated mostly with the Italian Dominican friar and philosopher Saint Thomas Aquinas, who developed the Greek idea and discussed it in his theological writings. Aquinas argued that the Universe works as sequences of cause and effect: the Universe exists and someone or something must have caused this. The cause is God; the effect is the world. Aquinas inferred that there must have been a First Cause. Nothing caused the First Cause – and the First Cause is God.

The argument remains philosophically controversial and most physicists do not find the argument convincing. Everything else they study that seems to be fundamental appears, on closer examination, to have an explanation lying behind it. Physicists are human and fallible, so they do not apply this principle consistently: historically, they called the basic constituents of gases 'atoms' (see page 18) as the first causes of chemistry, before discovering the components of atomic structure like electrons and protons, which explained some of their otherwise inexplicable properties. Physicists repeated the same mistake when they began to talk about 'fundamental' particles, which have been discovered not to be fundamental at all, but are made up of quarks and gluons. The range of fundamental particles and the number of their properties has turned out to be enormous, with little in the way of elegant explanations. The Higgs boson elementary particle is one explanation that has been developed in order to explain why the fundamental particles have the otherwise inexplicable masses that they do. It lurks in the background as a 'first cause' of particle physics, although no doubt its property of 'fundamental' will be assailed.

It is always going to be possible in cosmology to ask the question 'why?', and an answer to the question 'why did the Big Bang happen the way it did and create the Universe in the way it has been discovered to be?' is that the Universe only seems to favour our existence because we would not be able to inspect the Universe if we were not here. This is a truism but it falls short as a satisfactory explanation of what seems to be a significant fact.

Somewhat more subtle is the Multiverse theory: it suggests that there are many universes (the Multiverse), each with a different set of physical constants and laws. Of the many universes, few have constants with the values required to make our life possible, and of course we live in one of these select few. According to this viewpoint, this is why there are such favourable coincidences – the Universe has to be this way, otherwise we would not be here to observe that this is what it is like.

The Multiverse can be linked to inflation. Andrei Linde's theory of inflation proposes that, at the start of everything, quantum fluctuations cause tiny regions to expand rapidly and become isolated bubbles, one of which we inhabit as our Universe, living quarantined inside. This idea gives the possibility of the Multiverse a physical basis.

It is possible that there may be evidence on this topic in the image of the CMB. Inflation should impress a slight twisting pattern in the vibrations of the CMB's radio waves called B-mode polarization. It is a subtle effect and even the most accurate currently available measurements of the CMB by the Planck satellite do not establish its existence. Equipment specifically designed to detect it has been set up in the Atacama Desert (the POLARBEAR experiment) and in Antarctica (the BICEP experiment), and elsewhere.

An intriguing direct hint that we live in a Multiverse is the existence of an anomalously large cold spot 10 degrees in size in the CMB in the constellation of Eridanus. Is it significant or is it a chance blob? Given the statistical properties of the size and temperature of the other smaller, less cold spots in the image, the probability that the Eridanus cold spot could come about by chance is 1 in 50. These odds mean that it is not impossible that it is a fluky result, but they are long odds and suggest that more study might be rewarding and find something interesting. What might that be?

A significant but speculative explanation for the cold spot is, perhaps, that it is a defect that was caused by one of Linde's universes colliding with ours, like soap bubbles touching. This is a highly controversial explanation derived from a slightly less controversial theory, the Multiverse, and the unproven theory of cosmic inflation. The award of a Nobel Prize for the discovery of a second universe is probably not imminent.

The concept of the Multiverse might be attractive, yes, intriguing, yes – but it is speculative. However, the idea provides a possible escape route from the rather pessimistic conclusion of the last sentence of the previous chapter that our Universe ends in silent darkness. We can imagine another universe in the Multiverse that has the right physics, such that it is both open and populated with inhabitants. They might be able to be more upbeat about the far future of their universe than we are about ours.

Glossary

Abbreviations

ALMA   Atacama Large Millimeter/submillimeter Array

Caltech   California Institute of Technology

CCD   charge-coupled device

CMB   Cosmic Microwave Background radiation

ESA   European Space Agency

LIGO   Laser Interferometer Gravitational-Wave Observatory

eLISA   Laser Interferometer Space Antenna

M   Messier object – for example, M 31 (the Andromeda Galaxy)

NGC   New General Catalogue object – for example, NGC 1555 (Hind's Variable Nebula).

Units of measurement

Astronomical unit (AU) Distance from Earth to the Sun: 150 million kilometres (93 million miles).

Gauss A unit of magnetic field.

Joules A unit of energy.

Kelvin A unit of temperature, in size the same as a degree Celsius, but with the start of the scale at a temperature of absolute zero, about -273 degrees Celsius (-460 degrees Fahrenheit).

Light second The distance that light travels in 1 second: 300,000 kilometres (186,000 miles).

Light year The distance that light travels in 1 year: 9.5 million million kilometres (5.9 million million miles).

Mega-Hertz (MHz) A unit of the frequency of radio waves, equal to 1 million oscillations per second.

Micron One millionth of a metre, equivalent to one thousandth of a millimetre.

Solar mass The mass of the Sun: 2.0 million million million million million million kilograms (2.2 thousand million million million million short tons).

Terms

Accretion/accretion disc The process by which material falls onto a celestial body, pulled in by its force of gravity. Any swirling motion of the infalling material causes the material to orbit the celestial body in a flat plane called an accretion disc, prior to completing the infall.

Acoustic oscillations Oscillations of matter in periodic waves, like sound; in astronomy, such oscillations influenced the position of material flowing in the expansion of the *Universe* before the formation of galaxies and caused vestigial regularities in their present positions relative to one another.

Antiparticle Every type of *fundamental particle* is associated with an antiparticle having the same mass and opposite electric charge. Collections of antiparticles are called antimatter.

Asteroid A small, dark body or minor *planet* orbiting in a planetary system (in the solar system, usually orbiting between Mars and *Jupiter*). In origin, an asteroid may be a planet whose development has been arrested, or a fragment of a planet or another asteroid broken off by a collision. Small asteroids are termed *meteoroids*.

Atom The smallest unit of ordinary matter, the basic unit of the chemical *elements*; a *nucleus* orbited by *electrons*.

Aurora A luminous glow in the sky, usually over the polar regions, generated by the collision of solar *cosmic rays* with air.

Axion A conjectural elementary particle, possibly what *dark matter* is made of.

Big Bang The explosive event at the start of the *Universe*.

Binary star system Two *stars* in orbit around one another. Triple and quadruple star systems have three and four members, and so on.

Black hole A body of such high mass and small size that the force of gravity at its surface stops anything, even light, from leaving. *Supermassive black holes* are found in the centres of galaxies; *stellar black holes* are formed by *supernova* explosions; small black holes may have been formed in the Big Bang.

Comet A small body in the solar system like an *asteroid* but made principally of ice, which, warmed by the *Sun*, releases material in a tail.

Continent A large land mass, one of seven identified on present-day *Earth*.

Core The central, denser volume of material within a *planet, star* or *galaxy*.

Corona An outer halo of material that surrounds a *star* like the *Sun*; the Sun's atmosphere.

Coronal mass ejection A cloud of *plasma* thrown out of the solar *corona* into the solar system.

Cosmic Microwave Background Microwave (and infrared) *radiation* created in the *Big Bang*, pervading the whole of space as a uniform background seen in every direction.

Cosmic rays High-energy *ions*, originating in *stars* and elsewhere, and travelling at high speeds.

Crust The outer, solid, rocky layer of a *terrestrial planet*.

Dark energy A conjectured form of energy released from space that causes acceleration of the expansion of the *Universe*.

Dark matter A conjectured form of matter that emits no light, or any other *radiation*, but which exerts a gravitational force like the familiar, ordinary matter.

Dwarf planet A small *planet*. In the solar system, a planet sufficiently massive that it has taken up a spherical form due to the force of its own gravity, but not massive enough that it has attracted other material and cleared its orbital zone entirely.

Earth The third, rocky *planet* from the *Sun*.

Eclipse The occultation of a *star* (such as the *Sun*) by a *planet* or *moon*, or the shadowing of one planet or moon by another from the star that is the source of its illumination.

Electron An elementary particle, a constituent of *atoms*, negatively charged.

Element A pure chemical substance consisting only of *atoms* of the same type, all having the same number of *protons* in their nuclei. The larger the number of protons and *neutrons* together, the heavier the element.

Elliptical galaxy A *galaxy* with an overall elliptical shape, comprised principally of older, red *stars* and little or no gas.

Exoplanet A *planet* that lies outside the solar system, most typically in an exoplanetary system.

Exoplanetary system A group of *planets* in orbit around a *star* other than the *Sun*, together with associated smaller bodies like *asteroids*.

Fundamental particle A subatomic particle.

Galaxy A large collection of *stars*, interstellar gas, dust and *dark matter*, gravitationally bound together and isolated from other similar collections.

Gas giant planet A *planet* like *Jupiter* that is made, primarily, of gas; also known as a jupiter. Contrast with *terrestrial planet*.

Globular cluster A cluster of many *stars* having overall a spherical shape.

Gravitational wave A disturbance in spacetime caused by changes in motion of a celestial body that propagates with the same speed as light; a change of gravity in space that causes bodies to shift in position.

Half-life A measure of the time that it takes for a radioactive *element* to decay – specifically, the time that it takes for half of it to change to something else.

Hypernova A very energetic *supernova* in which the *core* of a massive *star* collapses to form a *black hole* with the ejection of the envelope of the star at high speed.

Inflation The conjectural period at the very beginning of the *Universe* during which space expands very rapidly by a very large factor.

Interferometer/ interferometry An instrument with multiple detectors that respond to a passing optical, radio or *gravitational wave* etc., with the output of all the detectors brought together and combined into a single response. Interferometry is the technique of using an interferometer.

Intergalactic medium The material that is distributed in intergalactic space (the space between galaxies).

Interstellar medium The material that is distributed in interstellar space.

Ion An *atom* with one or more *electrons* missing, or added, so that it is positively or negatively electrically charged.

Jupiter The largest *planet* in the solar system, the fifth planet outwards from the *Sun* and a *gas giant*. When spelt with lower case J, 'jupiter' means a planet (most often an *exoplanet*) that is massive like Jupiter.

Kilonova An explosion during which two *neutron stars*, or a neutron star and a *black hole* orbiting each other in a *binary star system*, merge into each other, producing *gravitational waves*.

Kuiper Belt The zone beyond Neptune where Trans-Neptunian Planets orbit, including Pluto.

Lithophile The chemical *elements* that bond with chemical elements in rocky material in a *terrestrial planet*'s *mantle*. Contrast with *siderophile*.

Magellanic Cloud One of two *galaxies* in orbit near our own Galaxy.

Magma Molten rock.

Magnetosphere The volume of space around a magnetic *planet* or *star*, permeated by a magnetic field.

Mantle The rocky zone lying between the *core* of a *terrestrial planet* and its *crust*.

Messier/NGC objects Catalogued objects in space.

Meteor A *meteoroid* that is falling onto a *planet* – for example, through *Earth*'s atmosphere, where it heats up and is visible as a luminous streak in the sky.

Meteorite A *meteoroid* that has fallen as a rock onto *Earth*.

Meteoroid A small *asteroid*, a rock or piece of dust orbiting in the solar system prior to falling onto *Earth*.

Milky Way Massed *stars* that create the impression of a band of light around the celestial sphere; our *Galaxy*, the shape of which causes this effect.

Molecule A combination of *atoms* that have bonded into a single chemical structure; the basic unit of a chemical compound.

Moon A satellite of a *planet*; the Moon is the satellite of our *Earth*.

Neutrino An elementary particle, electrically neutral with a very small, almost zero mass, and having a weak interaction with other particles. There are three flavours of neutrino, each associated with one of the following particles: *electrons*, muons and tau particles.

Neutrino oscillation The transition in flight of one flavour of *neutrino* to another.

Neutron An elementary particle found with *protons* in the *nucleus* of an *atom*, having almost the same mass as a proton but electrically neutral.

Neutron star A small, dense *star* that is made of *neutrons*.

Nova The explosion of a *star* so that a bright and apparently new star appears where no star was noticed before. See *supernova*, *kilonova*, *hypernova* for varieties.

Nuclear fusion The combination and amalgamation of atomic nuclei such as occurs in hot, dense matter in *stars*.

Nuclear reaction An interaction between nuclei that results in changes in the nuclei, releasing (or absorbing) energy.

Nucleus The central, heavy part of an *atom* that is usually orbited by the atom's quota of *electrons*. A nucleus is made up of approximately equal numbers of *protons* and *neutrons*.

Planet A body that orbits a *star* (or occasionally, is freely floating in space) and is too small to sustain *nuclear reactions*. See *terrestrial planet, gas giant planet, asteroid*.

Planetary nebula A nebula formed as a *red giant star* makes the transition to become a *white dwarf*. These nebulae do not have anything to do with *planets*, but look like planets when viewed through a small telescope, as was the case when they were first discovered and named.

Planetesimal A small, primitive *planet*; the stage in the growth of a planet when it lies somewhere in size between a rock and an *asteroid*. A *comet*.

Plasma The fourth state of matter (additional to solids, liquids and gas) and consisting of *ions* and *electrons*. The state of matter in hot bodies such as *stars*.

Proton A *fundamental particle*, positively charged with one unit of electricity and heavy, found in atomic nuclei. The *nucleus* of an ordinary hydrogen *atom*.

Protoplanet The last stage in the formation of a *planet*.

Protostar The last stage in the formation of a *star*.

Pulsar A rotating, magnetized neutron star, and a source of repetitive radio pulses.

Quark One of the six types of the most *fundamental particles* currently known, carrying an electric charge equal to either one-third or two-thirds of a unit.

Quasar An energetic *supermassive black hole* at the centre of a *galaxy*.

Quasi-Stellar Object See *quasar*.

Radiation Energy propagating through space.

Radioactivity The decay of the *nucleus* of an *atom* by the emission of *radiation*, changing the nucleus into another type.

Red giant/red supergiant A large/very large low-temperature *star*; a star at an advanced stage of development.

Redshift A shift in the colour of a *star* or *galaxy* caused by its motion as it recedes. Typically used of a galaxy as it partakes in the expansion of the *Universe*. The shift is the amount by which the wavelength of a spectral line of a particular colour is altered. If the line originates as a colour in the middle of a rainbow-like *spectrum*, the recessional speed of a galaxy shifts the colour away from blue and towards red, hence the origin of the term, which is used to describe an expansion speed, even for spectral lines that are not actually colours, but that are, for example, radio waves.

Resonance A condition of two orbiting celestial bodies – for example, *planets* whose orbital periods are in a whole number ratio, so that their configuration accurately repeats.

Siderophile An *element* that amalgamates with iron and sinks into the *core* of a *terrestrial planet*.

Solar system The *Sun*'s planetary system.

Spectrograph/spectroscopy A device that splits *radiation* into a *spectrum* of radiation of a progression of energy and records it in some way. Spectroscopy is the technique of using a spectrograph.

Spectrum *Radiation* arranged progressively in energy – for example, the rainbow of light that progresses from red, through orange, yellow and green, to blue.

Spiral galaxy A *galaxy* of *stars* and gas, with the bright stars and the gas laid out in a spiral pattern.

Star A large celestial body that generates and radiates energy (by *nuclear reactions*), which is held together by its own gravity and supported by an internal pressure.

Starburst A sudden surge of *star* formation in a *galaxy*.

Stellar black hole A *black hole* the mass of a star, produced by a *supernova* explosion.

String The underlying constituent of a *fundamental particle* that vibrates and interacts with other strings so as to make up the properties of the particle.

Sun The *star* at the centre of our solar system.

Supermassive black hole A *black hole* of mass in excess of, say, a million times the mass of our *Sun*.

Supernova An explosion of a *star* in which the outer layers of the star re-ejected and the *core* collapses to a *neutron star* or *black hole*, or entirely disintegrates.

Tectonics Geological processes in the *Earth*'s *crust*, particularly pertaining to the motion of continents.

Terrestrial planet A rocky planet like the *Earth*.

Universe The collection of everything.

Variable star (e.g., T Tauri, Cepheid) A *star* that varies in brightness, either because it actually does or because something passes in front of it. The various sorts of variable star are named after a typical member.

White dwarf A compact *star* produced as the exposed *core* of a relatively low-mass star.

WIMP A conjectural 'weakly interacting massive particle', possibly the particle that *dark matter* is made of.

Picture credits

Line illustrations on pp. 6, 9, 24, 47, 64, 88, 116, 141, 161, 181, 194, 215 and 249 by Amanda Smith

I ESO/B. Tafreshi (twanight.org)

II NASA, ESA, S. Beckwith, M. Stiavelli, A. Koekemoer (STScI), R. Thompson (University of Arizona), and the STScI HUDF Team, G. Illingworth, R. Bouwens (University of California, Santa Cruz), and the HUDF09 Team, R. Ellis (Caltech), R. McLure, J. Dunlop (University of Edinburgh), B. Robertson (University of Arizona), A. Koekemoer (STScI), and the HUDF12 Team, G. Illingworth, D. Magee, P. Oesch (University of California, Santa Cruz), R. Bouwens (Leiden University), and the HUDF09 Team, H. Teplitz, M. Rafelski (IPAC/Caltech), A. Koekemoer (STScI), R. Windhorst (Arizona State University), and Z. Levay (STScI)

III NASA, ESA, E. Jullo (JPL/LAM), P. Natarajan (Yale) and J-P. Kneib (LAM)

IV ESA/Planck Collaboration

V M. Blanton and the SDSS

VI ESA/Hubble & NASA

VII EHT Collaboration

VIII European Southern Observatory

IX NASA, ESA/Hubble and the Hubble Heritage Team

X NASA/SDO XI X-ray: NASA/CXC/ SAO; Optical: NASA/STScI; Infrared: NASA-JPL-Caltech

XII ESO/A. Müller et al.

XIII NASA/Johns Hopkins University Applied Physics Laboratory/Southwest Research Institute/Roman Tkachenko

XIV NASA/JPL-Caltech/SETI Institute XV NASA/JPL-Caltech/MSSS

XVI NASA; ESA; Z. Levay and R. van der Marel, STScI; T. Hallas, and A. Mellinger

XVII NASA/Bill Anders

Index